Praise for Oil Politics

Nnimmo Bassey embodies the thinker, writer, activist in one. His latest collection of essays *Oil Politics* is the story of our times. And since we are all eating, drinking, thinking oil, it is a story each of us should read. Oil has caused pollution in the Niger Delta and contributed to climate change. But it has also polluted democracy. As Nnimmo puts it, the story of oil is the story of 'The blind walk of autocrats in the vice grip of kleptocrats results in unrelenting pummelling of the grassroots.' We need to move from Oil to Soil, from Kleptocracy to Earth Democracy. *Oil Politics* is a call to action to each and every Earth Citizen.— **Dr VANDANA SHIVA**, philosopher, environmentalist, author, professional speaker, social activist

For decades, Nnimmo Bassey has been a relentless warrior against the ravages of the oil industry, holding the Niger Delta up as both a stark warning and an inspiring model of resistance. The truths in these essays demonstrate that the climate crisis amounts to a war, one waged by global elites on the poorest and most vulnerable. In his defiance, fearlessness and lyricism, Bassey also lights the way towards a just and democratic peace. — **NAOMI KLEIN**, author *This Changes Everything* and *The Shock Doctrine*

Nnimmo Bassey is that rare individual— he combines solid theoretical knowledge with practice; a perceptive writer and campaigner of the finest pedigree. In this collection of essays, ranging from issues of petroleum extraction to climate justice, Bassey brings to bear these formidable talents. This book deserves reading and re-reading. It is a worthy addition to the corpus of works on Africa's badly mauled ecology. — **Dr IKE OKONTA**, author *When Citizens Revolt: Nigerian Elites, Big Oil and the Ogoni Struggle for Self-Determination* and co-author *Where Vultures Feast: 40 years of Shell in Nigeria*

Very few people understand the 'politics of oil' and have confronted the environmental crisis in Nigeria like Nnimmo Bassey. In Oil Politics: Echoes of Ecological Wars, he not only reveals the devastating impact of our environmental indiscretions but how the incestuous relationship between the Nigerian state and multinationals like Shell has left Nigeria and Nigerians gasping for breath. If we still care about Nigeria, or what is left of it, then we can only ignore this intervention at our own risk! — **CHIDO ONUMAH** author, *We Are All Biafrans*

Oil and mineral development represents a continuous act of violence against nature and society; this violence is a prerequisite to these extractive activities. Faced with this reality, communities in diverse regions of the planet organize varied forms of resistance and construct alternatives. Nnimmo Bassey is one of the human beings most committed to ecological justice and thus, social justice. This book, a collection of the author's essays, is an example of that commitment. — **ALBERTO ACOSTA**, Economist, former President of the Constitutional Assembly of Ecuador, former Minister of Energy and Mines

Nnimmo Bassey is an angry good man, aware in his bones of the socio-ecological debt from North to South. He writes brilliantly calling the world to action for climate justice and against fossil fuels extraction. He comes from Nigeria and the Niger Delta where over two million barrels of oil are exported everyday, where many people have been killed while others have resisted throughout the decades of destruction brought by Shell and other companies.— **Professor JOAN MARTINEZ-ALIER**, ICTA, Universidad Autonoma de Barcelona

Nnimmo Bassey is one of the best known and most respected activist/analyst of the socio-political and environmental impact of fossil fuel extraction across the planet. As part of his commitments he has played a leading role in Friends of the Earth International, Environmental Rights Action in Nigeria and Oilwatch International. For more than two decades he has directly participated and/or documented peoples' struggles against these depredatory activities, not only in Nigeria, but also in South Africa, Equatorial Guinea, Ecuador, Brazil, the Gulf of Mexico and others. ... A main focus of his attention has been the struggles of the Ogoni people against the social and environmental devastating impacts of Shell's extractive activities in the Niger Delta. This book contains an extraordinary, thoughtful and well documented critical analysis of many of these impacts and struggles. The way in which multiple dimensions of the fossil fuel civilization are integrated into the analysis is particularly valuable: impact on people's lives; environmental devastation: climate change: the impunity with which transnational corporations operate in the Global South; government complacency and corruption; military repression; the geopolitics of oil; the implications and unsustainability of high consumption life styles based on cheap fossil energy; as well as the multiple forms of popular resistance and struggles. Activists and communities around the planet, who not only believe that another world is possible but are willing to fight for it, have much to learn from this book.— **EDGARDO LANDER**, retired professor of social sciences at the Universidad Central de Venezuela in Caracas, Caracas

OIL POLITICS

OIL POLITICS

ECHOES OF ECOLOGICAL WARS

NNIMMO BASSEY

,

Daraja Press

Published by

Daraja Press

www.darajapress.com

ISBN: 978-0-9952223-1-1

Cover design: Otoabasi Bassey

Library and Archives Canada Cataloguing in Publication
Bassey, Nnimmo, author
Oil politics : echoes of ecological wars / Nnimmo Bassey.
Issued in print and electronic formats.
ISBN 978-0-9952223-1-1 (paperback).–ISBN 978-0-9952223-4-2 (ebook)
1. Petroleum industry and trade–Environmental aspects–Nigeria–
Niger River Delta. 2. Petroleum industry and trade–Environmental
aspects. 3. Oil pollution of soils. 4. Oil pollution of water. 5. Liability for oil pollution
damages. 6. Social responsibility of business. I. Title.

TD195.P4B37 2016   338.2'728   C2016-906540-5

C2016-906541-3

Contents

Part VIII. AFTERWORD

List of Abbreviations

AHOMAR, Homens e Mulheres do Mar Association
API, American Petroleum Institute
ATS, Alien Tort Statute
BRICS, Brazil, Russia, India, China, South Africa
CDM, Clean Development Mechanism
CO2, Carbon dioxide
COP, Conference of Parties
CPC, Congress for Progressive Change
CPF, Central Oil and Gas Processing Facility
CSCEC, China State Construction Engineering Corporation
deg C, Degrees Celcius (Centigrade)
DPR, Directorate of Petroleum Resources
EIA, Environmental Impact assessment
EITI , Extractive Industries Transparency Initiative
EMP, Environmental Management Plan
ERA, Environmental Rights Action
FAAN,Federal Airports Authority of Nigeria
FAO, Food and Agriculture Organisation
FEPA, Federal Environmental Protection Agency
FPSO, Floating Production and Storage and Offloading
GDP, Gross Domestic Product
GON, Government of Nigeria
HNDC, Hope for Niger Delta Campaign
HYPREP, Hydrocarbon Pollution Restoration Project
IJV, Incorporated Joint Ventures
ILO, International Labour Organisation
INDC, Intended Nationally Determined Contributions
IPCC, Intergovernmental Panel on Climate Change
IYC, Ijaw Youth Congress
JTF, Joint Military Taskforce
LCA, Long-term Cooperative Action
LGA, Local Government Area
LNG, Liquid Natural Gas
LPG, Liquefied Petroleum Gas
MAN, Manufacturers Association of Nigeria
MEND, Movement for the Emancipation of the Niger Delta
MOSOP, Movement for the Survival of the Ogoni People
NAFDAC, National Agency for Food and Drug Administration and Control
NAP, National Action Programme
NAPEP, National Poverty Eradication Programme

NASS, National Assembly
NDDC, Niger Delta Development Commission
NDES, Niger Delta Environmental Survey
NDLF, Niger Delta Liberation Force
NEITI, Nigerian Extractive Industries Transparency Initiative
NESREA, National Environmental Standards and Regulations Enforcement Agency
NIALS, Nigerian Institute of Advanced Legal Studies
NNPC, Nigerian National Petroleum Company
NOSDRA, Nigerian National Oil Spill Detection and Response Agency
NRAN, No REDD in Africa Network
OECD, Organisation for Economic Co-operation and Development
OPEC, Organization of the Petroleum Exporting Countries
PANPP, Pan-African Non-Petroleum Producers Association
PDP, Peoples Democratic Party
PIB, Petroleum Industry Bill
REDD, Reducing Emissions from Deforestation and Forest Degradation
REDUC, Refineria Duque de Caxias
SADC, Southern Africa Development Community
SAGRC, South African Green Revolutionary Council
SEC, US Securities and Exchange Commission
SNEPCO, Shell Nigeria Exploration and Production Company
SON, Standards Organisation of Nigeria
SPDC, Shell Petroleum Development Company
SUDAPET, Sudan National Petroleum Corporation
TMDB, Transvaal and Delagoa Bay Mine
TNC, Transnational Corporation
UNCCD, United Nations Convention to Combat Desertification
UNEP, United Nations Environment Program
UNFCCC, United Nations Framework Convention on Climate Change
USAID, United States Agency for International Development
WHO, World Health Organisation

Foreword by Ogaga Ifowodo

In *Oil Politics: Echoes of Ecological Wars*, a collection of his essays, talks, blog posts and sundry reflections on the hot topic of the environment, Nnimmo Bassey takes a cue from a speech by Álvaro García Linera, Vice President of Bolivia, who restates Rosa Luxemburg's famous poser 'socialism or barbarism' as 'Mother Earth or barbarism'. In this important book, Bassey expounds the view that climate change is a product of the crisis of capitalism and 'its attendant creed of expansion and unlimited economic growth and profits,' adding that the current inability to confront the climate change crisis 'is due mainly to the vice grip on the global systems by the powerhouses of imperialism.'

I take Álvaro García Linera and Nnimmo Bassey further, perhaps, than they might or could have intended by wondering if the Christian variant of the belief in life after death to be lived in a heavenly domain, supposedly outside earth, has anything to do with the West's unwholesome attitude to climate change. 'This world is not my home, I'm just a-passing through,' sang the American folk musician Jim Reeves to the rapturous delight of millions. So the question must be posed to the Christian, capitalist West, is earth our home? The capitalist West holds itself out as the bastion of Christianity, but so preponderant is the evidence of its hugely disproportionate contribution to global warming that direct evidence is not needed. In short, in the court of environmental justice, circumstantial evidence alone would be enough to convict the West. If, however, you insist on hard facts, here's something to consider: although the United States of America constitutes a mere five percent of the world's population, it emits, nonetheless, about a quarter of the world's greenhouse gases. And together with Europe, is responsible for fifty percent of the non-naturally occurring carbon in the air, even though their combined population is only about ten percent of the world's inhabitants.

What, then, does it say about the western/capitalist attitude to the environment that in the United States, for instance, the ranks of the most strident deniers of climate change are too often filled with conservative Christians who have formed happy cause with Wall Street lords of mammon? Why have there not been loud and strident cries from the pulpits decrying the market mechanisms forced by the West on the rest of the world to ensure unrestrained environmental plunder, mechanisms that have crippled all the conferences of parties, from Copenhagen to Cancún and Durban, under the United Nations Framework Convention on Climate Change? These mechanisms—including the much trumpeted Reducing Emissions from Deforestation and Forest Degradation, a UN-sanctioned initiative—literally give carte blanche to corporate polluters to keep despoiling the environment but to salve their conscience by buying 'carbon credits.' What are 'carbon credits?' many will ask—as I did until Bassey explained the dubious term to me in *Oil Politics*. It is the asinine, but profit-grabbing, idea that a big corporation, say ExxonMobil or Shell, can pollute the immediate environment of any of its undertakings—an oil field or a refinery, for example—to its heart's content but then to offset the looming danger through a purported 'green' project elsewhere, say a vast mono-culture plantation, almost always in the Third World. In its more cynical form, it is the unvarnished view that the alarming rate of pollution of the highly industrialised and consumerist West—which, in its strictly economic sense includes China, 'the factory of the world' and other titans of the capitalist plundering of the earth—is proportionately offset by the vast stock of carbon credits in the rest of the un- and under-developed world in Africa, South and Central America and Asia.

What might happen to the economic ethos of the more polluting West if it embraced the not-so-esoteric idea that this earth is, after all, our home? In April 2010, following the predictable failure of the 2009 Copenhagen Conference Of Parties (COP) which ended without an agreement—the government of Bolivia hosted the more forward-looking Cochabamba *World Peoples Conference on Climate Change and the Rights of Mother Earth*. For the concept of Mother Earth, as Álvaro García explains 'is not just a slogan. It means a new way of producing, a new way of relationship with nature and with one another. This relationship is one of equality and not domination, a relationship of dialogue, of giving and receiving. It is not merely a philosophy or folklore.'

Bassey is in the front lines of the earth-is-our-home ethos. *Oil Politics* ought to be read by all concerned about the unsustainable path of mindless corporate greed, its counterpart in thoughtless consumption, and the stiff price of both paid by our earth. The environment can seem, ironically, too intimate and yet too far-fetched all at once. We live in it, yet its immanence, its seeming infinity, makes it sound alarmist to speak of human activity bringing it to an apocalyptic end (though billions have no problem believing in an apocalyptic end caused by sin). The oceans rising and swallowing up coastal countries? Bah! Deserts creeping on savannahs and turning loamy plains into parched acres of sand? Thank you, but over here where we are, our problem is flooding caused by endless rains! Always, we tend to misread the signs or deny them outright because they do not accord with our immediate experience.

Bassey explains in lay terms, often with humour though each topic he tackles be ever so grim, some of the esoteric terms in which climate talk is often discussed such as: 'greenhouse effect', 'carbon credits', 'carbon capture', 'global warming', 'market mechanisms' and what a two degree Celsius rise in temperatures would mean for tropical Africa as opposed to Greenland, or a one metre rise in the sea level would to littoral communities in Nigeria. At about the same time that Bassey served out his term as president of Friends of the Earth International, and passed on the torch of leadership of Environmental Rights Action, Nigeria's premier nature NGO, he set up Health of Mother

Earth Foundation, through which he currently does most of his advocacy. Yes, the earth is his home and ours too.

A moment like this calls for some story-telling. I first met Bassey as a sophomore in law at the University of Benin where he was the principal architect. Bassey would later lead the university's physical planning team to design the temporary administration building and the vice-chancellor's lodge. That was during the turbulent period of General Babangida's invasion of the universities in search of radicals and extremists. The vice-chancellor, Professor Grace Alele-Williams, gladly delivered several scalps, among them those of Dr Festus Iyayi, president of the Academic Staff Union of Universities at the time, and Professor Itse Sagay, dean of law. This was meant as dire warning to all outspoken members of the increasingly restive universities, whether on the faculty or in the administration. Bassey, a regular on the op-ed pages of *The Guardian* and *Vanguard*, became visibly disenchanted—or 'disgruntled' in the power-speak of the time—and soon enough quit the Ivory Sewer to set up *Base Consult*, his private architectural practice, just outside the university. And it is here that the idea of a human rights approach to the defence of the environment took seed, bloomed and flourishes now to worldwide acclaim.

The Environmental Rights Action (ERA), partner group of global watchdogs Friends of the Earth and OilWatch, began as a project of the Civil Liberties Organisation (CLO) where Bassey was a member of the governing board and chairman of its southern zone. This project yielded several shocking exposés on the mindless devastation of the environment by oil and timber companies such as Shell in Iko (Akwa Ibom) and WEMCO at Omo Forest (Ondo), to cite only two examples. The effect was as crucial for the emergence of an environmental consciousness in Nigeria as CLO's uncovering in 1987 of the horrors of Ita-Oko, an off-shore prison built under General Obasanjo as military head of state, as it was for the emergence of consciousness of civil liberties. It was soon clear that ERA needed a wider field of vision than the CLO's fulcrum of political liberties would allow. Not even the compromise of semi-autonomy marked by the new name, ERA/CLO, would widen this field by the needed acreage. The debate was often passionate, for ERA had come to deepen CLO's work in a way that added to its prestige at home and abroad, but CLO voted wisely to let the child strike out and come into his own.

This decision freed ERA under Bassey to synthesise its mission from indigenous principles of harmonious natural resource management, Christian teachings on egalitarianism and the international declaration of human rights. The result is a charter of struggle for the attainment of social justice. Driven like only a pastor—in the true sense of herdsman, shepherd—can be about his flock, it is no surprise that Bassey's e-mail signature is a plea, borrowed from the great Psalmist, to the principalities and powers of earth on behalf of the oppressed, exploited and downtrodden: 'How long will you keep judging and favouring evil people? Be fair to the poor and orphans. Defend the helpless and everyone in need. Rescue the weak and homeless from the powerful hands of heartless people.' These are the slogans that inform Bassey's vision of a habitable and sustainable earth. And for his steadfastness, he was driven underground for four months in 1994 as his brother-in-law was held hostage, arrested and detained by the secret police, the Nigerian State Security Service (SSS) for several days in October 1997 and had his passport impounded for long periods.

Bassey, who has published four collections of poems, including the grim title, *We Thought It Was Oil But It Was Blood*, has also been General Secretary of the Association of Nigerian Authors. Let us pay heed to the Reverend Nnimmo Bassey, a hero of the environment, and, even more, be ourselves in small and big ways architects of a sustainable earth.

September 2016

Preface

The essays preserved in this collection were written as responses to the unrelenting ecological assault on local communities in Nigeria, across Africa and elsewhere. Some speak of the unyielding wedlock between governments and transnational corporations in what would probably best illustrate what is meant when people claim that love is blind. The blind walk of autocrats in the vice grip of kleptocrats results in unrelenting pummelling of the grassroots.

Most of the essays contained in this volume were written as short opinion articles for Nigerian newspapers between 2010 and 2013. A number of these were published in a weekly column titled Oil Politics that ran in the then promising Nigerian newspaper 234NEXT that suddenly went kaput. The online links no longer lead to the articles. I now post newer articles as blogs[1]. Hopefully these will be better preserved for the future.[2]

The longer pieces in this collection are based on speaking notes and talks presented at meetings and conferences. As would be expected, they were responses to issues that either were in the news at that time, or in some cases issues that did not make it into the news.

The essays here largely deal with on petroleum extraction and climate justice. They are mostly focussed on Nigeria's Niger Delta – one of the most polluted places on the planet. It is instructive that although the ecocide visited on this region by the oil companies is well acknowledged, only perfunctory mention is made of their crimes in the media.

The reader may wonder why these dated essays have been collected together in this book. The fact is that the issues that I consider here are unresolved and continuous—what happened decades ago is still happening today. The demands of local communities for environmental remediation and the halting of destructive extraction remain unheeded today as they were then. The international corporations still operate behind military shields. Demands for social inclusion by local communities are still met with brute force. While

---

1. See at www.nnimmobassey.net or www.nnimmo.blogspot.com

2. EDITOR'S COMMENT: For those essays published elsewhere, we have done some light editiing for consistency as well as to correct errors.

progress has been made on some fronts as a result of determined vigilance of the suffering communities, there remains much to be done.

While a larger part of the blame for the destruction of the environment must lie at the door of the multinational corporations, it is certainly the case that there are individuals in communities who are also responsible for polluting for personal gain to the detriment of the majority of the people. Examples of such activities are the highly polluting bush refineries of the Niger Delta.

The tragedy is that environmental laws are not in short supply, neither in Nigeria nor elsewhere. But the problem lies in the failure of enforcement of such laws and regulations. Polluters are known to ignore court or executive orders, seeing themselves as being above the law and above sanctions. We see this repeatedly in Nigeria. Illegal activities, such as gas-flaring, continue to be perpetuated. There are minuscule penalties for infraction. In some cases, proposed legislation is 'booby-trapped' so that either it never sees the light of day or incorporates such huge loopholes that crudest criminals can casually walk through with impunity.

The same can be said of the UN climate change negotiations where we have witnessed systematic lowering of ambition and a disavowal of commitments as each year goes by and as the problems become more entrenched. I highlight the fact that Africa bears disproportionate impacts from a problem that she did not create. Some of these burdens are heaped on her by way of false solutions such as the Reducing Emissions from Deforestation and Forest Degradation (REDD).

The essays here underscore the vital need for individuals and communities to take more than a passing interest in the Conference of Parties (COP) and United Nations Framework Convention on Climate Change (UNFCCC). The lack of ambition evidenced in these negotiations and the fact that the aggregate outcomes are not taking us out of the woods but digging deeper into inescapable holes should wake us all up to action. The planet is facing destruction both as a result of the lack of action and as a result of the adoption of 'false solutions' that only exacerbate the speed of climate change. But much of what is agreed is frequently worded in inaccessible jargon. We face the challenge of sharing our views, stories and experiences in popular language to ensure that everyone can participate in developing political alternatives to the rhetoric of these conferences.

Cultural awakening was a tool that entrenched the Ogoni struggle and makes it stand out as one of the most successful mobilisations of heavily deprived peoples. The struggles of Ken Saro-Wiwa and the Movement for the Survival of Ogoni People (MOSOP) echo in some of the essays and indicate how unresolved environmental issues continue to persist over decades. The fact that the struggles are documented and preserved in various media helps to keep the window for action open. In the face of dastardly degradation, silence is not an option. Or, as Saro-Wiwa put it, 'Silence would be treason'[3]. Non-violent resistance remains the way in which those communities who have been most affected around the world continue to advance their cause. And that is what builds hope and shows that even where the sky is dark victory can still be snatched from the jaws of cannibal polluters.

The essays that are preserved in this collection will contribute, I hope, to developing and deepening an understanding of the ecological challenges ravaging Nigeria, Africa and our world today. They illustrate the global nature of these terrors. These essays are not meant just to enable coffee table chatter. No. These essays are intended as calls to action, as a means of encouraging others facing similar threats to share their experiences.

3. See: http://darajapress.com/catalog/silence-would-be-treason-last-writings-of-ken-saro-wiwa (accessed 26 September 2016)

PART I

ECHOES OF AN ECOLOGICAL WAR

1

I will not dance to your beat

I will not dance to your beat[1]
If you call plantations forests
I will not sing with you
If you privatise my water
I will confront you with my fists
If climate change means death to me but business to you
I will expose your evil greed
If you don't leave crude oil in the soil
Coal in the hole and tar sands in the land
I will confront and denounce you
If you insist on carbon offsetting and other do-nothing false solutions
I will make you see red
If you keep talking of REDD and push forest communities away from
their land
I will drag you to the Climate Tribunal
If you pile up ecological debt
& refuse to pay your climate debt
I will make you drink your own medicine
If you endorse genetically modified crops
And throw dust into the skies to mask the sun
I will not dance to your beat
Unless we walk the sustainable path
And accept real solutions & respect Mother Earth
Unless you do
I will not &
We will not dance to your beat

---

1. Title poem in Nnimmo Bassey's collection *I will Not Dance to Your Beat*, Ibadan, Kraft Books, 2011

2

Echoes of an ecological war

This article was written to mark the 15th anniversary of the murder of Ken Saro-Wiwa and eight other Ogoni leaders. It was published on 14 November 2010[1]

The world's addiction to fossil fuels put the hangman's noose around the neck of Ken Saro-Wiwa and eight other Ogoni leaders on 10 November 1995. That noose was tightened under the watch of Shell through a kangaroo military tribunal rigged by the worst dictator Nigeria ever had. Today, we can say that every oil rig that sucks oil in the Niger Delta is a hangman's noose around the necks of the suffering peoples and communities.

Today, we all stand before history. We stand in front of a backdrop of injustice, oppression, and ecological genocide—not just historical, but current and it is the threat of its progressing into the future that we must stand together to fight.

In his statement after the verdict of guilt was passed on him, Ken Saro-Wiwa declared, 'We all stand before history... appalled by the denigrating poverty of peoples who live in richly endowed lands.[2]

We stand distressed by 'their political marginalization and economic strangulation, angered by the devastation of their land and their ultimate heritage.'

He went on to call for 'a fair and just democratic system which protects everyone and every ethnic group, and gives us all a valid claim to human validation.'

Ken Saro-Wiwa's words, though spoken fifteen years ago, still ring true in our ears today. A man with a keen sense of history, he told the agents of the military dictator that he and his colleagues were not the only ones on trial. Hear him: 'Shell is here on trial, and it is as well that it is represented by counsel said to be holding a watching brief... The company has, indeed, ducked this particular trial, but its day will surely come, and the lessons learnt here may prove useful to it, for there is no doubt in my mind that the ecological war that

1. http://nigeriang.com/money/nnimmo-bassey-echoes-of-an-ecological-war-2/5527/ (accessed 7 June 2016)

2. Ike Okonta and Oronto Douglas. 2001. *Where Vultures Feast: Shell, Human Rights, and Oil in the Niger Delta*. Sierra Club Books, New York. P208

the company has waged in the Delta will be called to question sooner than later and the crimes of that war be duly punished. The crime of the company's dirty wars against the Ogoni people will also be punished.'

A man of history

Saro-Wiwa was indeed a man of history. While shackled in one military jail or another, the world recognised his worth and the validity of the Ogoni struggles. In the last months of his life on earth, he won several awards in recognition of his just struggles: the Fonlon-Nichols Award for excellence in creative writing and the struggle for human rights; the 1994 Right Livelihood Award or Alternative Nobel Prize for Peace; the 1995 Goldman Environmental Prize, the most prestigious environmental award in the world; the 1995 Bruno Kreisky Prize for Services to Human Rights; the 1995 British Environmental ad Media Special Awareness Award; and the Hammett Award for Human Rights of Human Rights Watch.

The Students Union of the Ahmadu Bello University in Nigeria conferred on him the award of Grand Commander of the Oppressed Masses. Surely, none of these could have been given to a man of mean repute.

Standing on the shoulders of history, we see clearly the beginnings of the trials that were bound to expose those who have waged ecological wars against the Ogoni people, the peoples of the Niger Delta and elsewhere in the world. We continue to see a company like Shell bowing before courts and before the Stock Exchanges in North America, accepting out of court settlements, and paying fines to avoid prosecution on bribery and corruption charges.

In 2005, they also admitted to having falsified their crude oil reserve figures the previous year, and paid some hefty fines to cover that up. Recent reports have it that they are halting suits over bribery by paying some fines. Last year, they agreed in a New York court to pay over US$15 million to Ogoni litigants for human rights abuses. [3]

In all these, we are confident that the words of Ken Saro-Wiwa will come to pass. One day, the eco devourers will have their day in the dock. And this is already happening in The Hague, where three Niger Delta communities are suing Shell for environmental degradation.

Perpetual death

The dominant predatory production and consumption patterns in the world, and the myth that crude oil is a cheap form of energy, has meant perpetual death sentence on communities where there is crude oil and gas.

If good men like Ken Saro-Wiwa had stayed silent[4] and allowed the

---

3. Shell Agrees to Pay USm for Role in Murder of Ken Saro-Wiwa, Other Ogoni Leaders. http://www.skycoded.com.ng/forum/8362-shell-agrees-to-pay-15m-for-role-in-murder-of-ken-saro-wiwa-other-ogoni-leaders (accessed 7 June 2016)

pattern of environmental degradation by oil extractive activities to go on unchallenged in Ogoni land, it is conceivable that things would have been worse by now.

Today, on account of the massive oil spills, gas flares, and careless handling of other industry-related toxic pollutants, life expectancy in the Niger Delta has plummeted to 41 years. If Ken Saro-Wiwa had not started the struggle, perhaps life expectancy would have possibly nose-dived to 20 years.

We stand before history and affirm that a sane future must be built on the platform of solidarity, dignity, and respect for the rights of Mother Earth.

We demand an end to fossil fuel addiction: be it crude oil, tar sands, or coal. We call for a Sabbath of rest for Mother Earth. Over the years, she has been abused, raped, and exploited and it is time to say enough is enough.

The blood of Ken Saro-Wiwa and all those massacred in the ecological wars for crude oil cry out today in demand for remaining oil to be left in the soil. With less than 40 percent of crude oil still left in the soil, it is foolishness to insist that we can go on driving on dregs through eternity....

---

4. A collection of the last writings of Ken Saro-Wiwa was published posthumously under the title **Silence Would be Treason.** 2013. Daraja Press

3

Human rights and the multiple environmental changes

Paper presented as the 2012 Rafto Prize Laureate at Rafto Conference, Bergen, Norway, 3 November 2012
A close look at the state of planet earth today shows a bleak picture and demands that humankind makes radical choices as to whether to further darken the future or to light some lamps and offer rays of hope.

The world is gripped in the embrace of multiple crises including those of climate, energy, food, water, finance and economy. These are all intensified by the reign of speculation on a path of unending growth based on the unspoken supposition that planetary space and resources are elastic and inexhaustible. This vision of the world is built on the premise of profit before people and on the creed that nature must be commodified in order to have *value* and stand a chance of being protected and defended. A creed aptly captured in the so-called 'Green Economy' concept in which everything carries a monetary price tag.

We are seeing a situation where market mechanisms are advertised as saviours of the world and the propping up of failing markets, corporations and financial institutions is set as the topmost job for politicians and policy makers. Still failing and failed markets, failed corporations and business moguls litter the world, literally. The burden of the failures and reckless exploitation falls on the shoulders of the impoverished and the oppressed who swell the ranks of the jobless, the homeless and the hungry.

The food crisis refuses to go away because food prices are driven by speculation. With the help of a few global food manufacturers the poor get fed plastic or junk foods and forests get cut down to raise cattle to meet unending demands for meat. Crops are grown for machines and lands are being grabbed under the guise of investment.

Starting from the erection of market mechanisms at Kyoto as the key way to tackle climate change, the stage was set for foisting false solutions through schemes like the very upsetting carbon offsetting and carbon trading ones. In order to build up the carbon market, trading systems have been created, but the price of hot air (carbon) has simply refused to fly. Biological carbon sequestration continues to capture the imagination of governments and traders, as does the pursuit of the yet to materialise mechanical carbon capture and storage. Seeing trees as nothing but carbon stocks and plantations as forests raise deforestation to new heights.

For a time there were arguments that agro-fuels could replace fossil fuels and would be good for the climate. The idea of ramping up agro-fuels/biofuels was principally a way to sustain the fossil fuel infrastructure since the production and utilisation paradigm are essentially the same. In other words, biofuels help keep fossil fuels civilisation on life support. Sadly for proponents, biofuel is not the energy solution. Indeed biofuels have since been seen to be a pipe-dream, although not before the notion set in motion an inexorable scramble for lands in the global south that has been aptly described as the great land grab. It is no longer news that the world has reached peak fish, peak oil and peak water. These are all signs of deep environmental changes, of more crises to come and of more human rights abuses as the Darwinian struggle for diminishing resources intensifies.

The series of changes driven by the multiple crises present a clear picture of environmental change. This concept recognises the fact that the environment is impacted by physical, economic, socio-political and other changes. We already know how closely tied the economic and political spheres are and how they collude to ensure that the environment and the people are exploited without redress or responsibility. The massacre of miners at Lonmin, South Africa highlights the extreme case of disposability of labour. Another manifestation of this situation is found in the sweatshops of Asia and elsewhere where workers are treated worse than pieces of machinery and are driven relentlessly by taskmasters. In this situation, the poor do not matter either because they do not have enough disposable income to make them desired players in the markets.

Desperate extraction

Fossil fuels have driven current modes of civilisation for over one and a half centuries. Coal, crude oil and gas enabled the world to shift from humans and animals as energy generators to machines that opened the highway to endless consumption. Crude oil appears cheap because the real costs are externalised. Today, with the days of easy oil ending, we are seeing a push into extraction in deep waters and fragile ecosystems. Some of the fragile ecosystems already being drilled include those in the Rift Valley of East Africa where oil and gas are being exploited in pristine environments and nature reserves. Other areas that are being sought after by the extractive sector and partner politicians include the Arctic region (where melting glaciers are seen as an opportunity and not an alarm), Yasuni ITT in Ecuador and offshore Lofoten in Norway. These and similar should be clearly off limits to polluting activities.

The end of easy oil is equally driving the increased adventures into fracking and deadly extraction of tar sands in Canada and elsewhere. While fracking is increasing domestic supply of oil and gas in the USA, tar sands increase fossil fuel exports from Canada; they are equally bringing up higher levels of environmental degradation with attendant health impacts that clearly impinge on human rights of citizens.

At this juncture, it is evident that humanity needs to get off the fossil fuels anaesthesia to be able to see that the extractive logic is simply not the way to keep a development path that has gone bankrupt. Consumption and endless growth present the dilemma of systemic greed overtaking inherent human greed and desire for accumulation of resources. Endless growth does not recognise that nature has boundaries and requires huge spans of time to replenish depleted resources. The intrinsic immorality and inequities in this situation seek ways of staying alive through the privatisation of conscience via media, particularly through electronic advertisements. All these actions are relentlessly pursued to appease or assuage corporate schizophrenia. It is time for rapid investment in truly renewable energy.

I have a dream. I have the dream that one day, offshore oil platforms and floating stations will become wind and solar farms. I have a dream.

The impunity of oil spills

Coming from a country where there is an equivalent of one Exxon Valdez volume of crude oil spewed into our environment yearly, it is inescapably clear that the petroleum industry is a very polluting sector. According to Senator Saraki, chair of Nigerian Senate's committee on environment, 'Oil spillage is not an oil business it is an environmental problem. Oil spill is an irresponsible environmental behaviour. The fact that it is as a result of oil exploration does not detract from the impact on the environment. Nigeria has lost over 13 million barrels of oil to preventable spills.' Senator Saraki added,

> It has been acknowledged by several reports including the UNEP Report that fifty per cent (50 percent) of oil spills in Nigeria has been due to corrosion of oil infrastructure, twenty eight per cent (28 percent) to sabotage and twenty one per cent (21 percent) to oil production operations. One per cent (1 percent) of oil spills is due to engineering drills, inability to effectively control oil wells, failure of machines, and inadequate care in loading and unloading oil vessels. It is the responsibility of the spiller to rehabilitate oil spill sites. It is as simple as that. The number of identified sites is over 2,000. The majority of these sites are sites with identified spillers. This gives an indication of the problem we already have in our hands.

It is obvious that there cannot be this level of ecological impunity without human rights being consistently trampled on. One quote from a Shell general manager in Nigeria in 1995 underscores the fact that impunity is good for some business: 'For a commercial company trying to make investments, you need a stable environment...Dictatorships can give you that.' This statement was made in early 1995 and by November Ken Saro-Wiwa and eight other Ogoni compatriots were hung by the dictatorship in power in Nigeria at that time.

Earlier in 1990, when the community of Umuechem protested against Shell's oil operations, Shell sent an urgent request for government security protection, requesting the 'Mobile Police' – units well-known for their brutality. The result was a two-day wave of violence that left 80 people were dead and nearly 500 houses destroyed.

Umuechem heralded a reign of terror that was visited upon the Ogoni people when, a few years later, they rose up in protest against oil operations that had resulted in minuscule local benefits but tremendous environmental costs. Again Shell relied on Nigerian security forces to secure its operations. Hundreds of Ogonis were arrested, tortured, and killed.

Efforts to obtain justice have taken impacted Nigerians to courts in Europe and USA. There is the case of four farmers and fishermen suing Shell in The Netherlands over pollution in Nigeria. Judgement is expected on 30 January 2013 in that case.

In 2002, a group of Nigerian plaintiffs brought suit under the ATS in a US federal court against a Shell's parent company, Royal Dutch Petroleum, for assisting in extrajudicial killings, torture, and crimes against humanity against the Ogoni people. These plaintiffs were living in the United States because they had received asylum from the US government due to their persecution in Nigeria. On February 28, 2012, the case, Kiobel v. Royal Dutch Petroleum (Shell), was argued before the US Supreme Court. Since then, the Supreme Court has ordered a second round of arguments, which took take place on 1 October 2012. This case is currently before the US Supreme Court with Shell launching a critical attack on human rights protections before the court by trying to gut a 200-year-old American law called the Alien Tort Statute (ATS). This law was originally used to bring cases against pirates but has developed into a way to bring suit against individuals and corporations that commit the worst types of human rights abuses like genocide, torture, and crimes against humanity.[1]

The following information from the website of Center for Constitutional Rights (CCR), the organization that supported the Kiobel case reveals the idea behind the case and how it all evolved:

> This lawsuit was brought under the Alien Tort Statute[2] and is a companion case to three lawsuits brought by CCR in 1996 against Royal Dutch Petroleum Company (Shell) for its complicity in human rights abuses against the Ogoni people in Nigeria.[3] After the Second Circuit Court of Appeals decided in 2000 that the court

1. The US Supreme Court eventually heard two rounds of arguments on the case on 28 February and 1 October 2012. The court's decision delivered on 17 April 2013 dismissed the case and affirmed the lower court's earlier decision.
2. See http://www.ccrjustice.org/sites/default/files/assets/files/ATS%20factsheet%209.2012.pdf

had personal jurisdiction over the Shell parent companies in CCR's case Wiwa   et   al   v.   Royal   Dutch   Petroleum, the Wiwa and Kiobel cases were consolidated for discovery, but because the cases had different procedural postures,Kiobel was appealed to the Second Circuit while Wiwa was set for trial. The Wiwa cases were settled on the eve of trial in 2009, providing a total of $15.5 million to compensate our clients, establish a trust for the benefit of the Ogoni people, and cover some of the legal costs and fees associated with the case. The Kiobel case, on the other hand, ended up heading to the Supreme Court on the question of whether corporations can be held accountable for human rights abuses and whether U.S. federal courts can hear claims arising from human rights violations committed abroad under the ATS.

For decades, CCR has pioneered the Alien Tort Statute as a tool to pursue international human rights violations in U.S. courts. Thus, we actively supported the plaintiffs in Kiobel as their case headed to the Supreme Court on questions regarding the ATS. CCR and allies filed numerous amicus briefs insupport of the Kiobel case against Shell, three of which were filed in the Supreme Court. In a 2007 amicus brief, filed on behalf of the Wiwa plaintiffs, CCR argued that extra-judicial killing is an actionable norm under the ATS.

On April 17, 2013, the Supreme Court issued its decision in the case, ruling that the Alien Tort Statute could not be applied to Shell's actions in Nigeria. The Court's decision undercut 30 years of jurisprudence to limit U.S. courts' ability to hear cases on human rights violations committed outside the U.S, limiting the ATS to those cases that "touch and concern" the U.S. with "sufficient force." CCR has continued to develop ATS jurisprudence in other cases post-Kiobel. We remain fully committed to continue challenging corporate human rights abuses and abuses by individual torturers and war criminals, no matter where they are committed.

The oil company's arguments were interesting: They argue that US law should not allow holding companies responsible for committing the most severe atrocities. They also claim that domestic US courts have no business in holding multinational corporations responsible for human rights abuses, especially those that happen in other countries. But as a 2007 UN report confirms, numerous countries recognise corporate legal responsibility for violations of international law.

If the Supreme Court does what Shell's asking it to do — grant immunity for human rights abuses committed overseas — this would allow mega-

---

3. See Wiwa et al v. Royal Dutch Petroleum et al. at http://ccrjustice.org/home/what-we-do/our-cases/wiwa-et-al-v-royal-dutch-petroleum-et-al for more information

corporations to operate by a different set of rules around the world and would turn the clock back more than 200 years.

The rain that beats us all

The impacts of global warming are clear and some of these include: intensified desertification, changed rainfall patterns leading to unusual floods, sea level rise and increased coastal erosion. These directly impact fragile infrastructure as well as the capacity of vulnerable peoples to reap expected fruits of their labour in the area of agriculture due to both loss of arable lands and the increasing salinity of coastal water bodies.

The fundamental causes of climate change are well known and documented. The basic drivers are anthropogenic actions, especially the use of fossil fuels for energy production resulting in the release of huge quantities of greenhouse gases into the atmosphere. It is important that we note the fact that while the bulk of the polluting activities have occurred and continue to occur in rich industrialised world, the impacts are mostly felt in nations and regions contributing least to the problem. We note that 50 per cent of the carbon in the atmosphere has come from just the USA and the European countries whose populations add up to only 10 per cent of the world's population.

Stressing the fact of fossil fuels being the fundamental culprit in this crisis, Oilwatch International regrets that petro-independence is being strengthened by the day.

Overall energy consumption grew by 5.6 percent in 2010, with continued dependence on fossil fuels (coal, gas and oil). This was the highest rate of growth since 1973 and energy consumption is growing in every region of the world, and particularly in China.

Petro-dependence is maintained through private sector and state strategies that include both direct violence and the indirect violence of advertising bombardment, greenwashing and political corruption. In spite of local opposition, environmental impacts and economic illegalities, and in spite of the general crisis to which it is linked, this strategy is maintained and even imposed as a priority.

Small island nations and other vulnerable nations in Africa, Asia and Latin America are in the frontlines of areas that are most threatened by climate impacts. Sadly, these nations are forced to focus their energies on planning and taking actions to adapt to the changing scenarios foisted on them. They are also forced to seek ways of mitigating the impacts of climate change.

Systemic challenges, systemic solutions

The United Nations Framework Convention on Climate Change (UNFCCC) was set up to provide guideposts and to urge nations to act together in everyone's common interest: for the survival of the planet as we know it. The conferences of parties (COP) to the convention have quickly turned into talk shops for nations to display their power, arm-twist the poor and evade action. This became most obvious at COP15 in Copenhagen and became entrenched at COP16 in Cancún. What happened at COP17 in Durban must take the medal as a conference whose critical achievement was the blatant postponement of action while the earth burns.

Climate justice activists saw the Durban debacle as one where ordinary people were unabashedly let down by governments. There are no reasons why we should be optimistic about the outcome of the next Conference of Parties that will be held in Qatar.

The demand of climate justice is that those who created the climate problem must be the ones to mitigate it. There are two ways to go about it. First, rich nations must reduce their rapacious consumption patterns and address the climate crisis with real solutions and not ones that have been seen to be false. Second, the rich nations have to support the poor nations who are being forced to adapt to a situation they did not create. One way of practically making that to happen is through the support for sustainable low-carbon development paths.

At the close of the Durban meeting the descent into non-binding pledge-and-review system was set. This system of inaction places the world on the way to warming by as much as 4 to 7 degrees Celsius above preindustrial levels. If that happens, Africa will be cooked because the continent experiences 50 per cent more than average global temperature levels. The reluctance of rich and highly polluting nations to take real action on climate change has rightly been described as a form of apartheid. This is apartheid against Mother Earth and the species that she bears.

Love in Kyoto

It is worth recollecting at this juncture that the key point of the Kyoto Protocol is that it has legally binding emissions reduction requirements for industrialised or Annex 1 countries. In reality the protocol set very minimal targets for reduction of carbon emissions that were to be achieved between 1990 and 2012. Major emitters such as the USA and Australia did not accept these targets. The UNFCCC and other analysts have shown that even if the targets set by Kyoto were met, the climate crisis would not have been sufficiently tackled.

One of the key failures of the Kyoto protocol is that it did not unambiguously pin the blame for the problem on hydrocarbons. As long as this was the case, the frameworks for handling the problem were fundamentally flawed. Conventional wisdom instructs us to tackle the root causes of problems rather than the symptoms if we wish to radically pursue long lasting solutions.

Weak as it was, the Kyoto Protocol itself was not adopted easily and this delay was largely due to the withdrawal of the USA, the global giant in carbon emission, in 2001. It is important to note that before the USA withdrew they had effectively influenced the language of the protocol and firmly planted the bent to carbon mercantilism or 'free market' environmentalism. In fact they got the world to accept the market language and concept at the 1997 Kyoto meeting in exchange for USA support that never materialised. The world is still stuck with the mindset of these untested ideas to this day. The protocol was set on a market ideology and this has blocked the pathway to real and just solutions to climate change.

We note here that although the USA has a mere five percent of the world's population it emits nearly 25 percent of the world's greenhouse gases from the burning of oil, gas, and coal – for driving cars, producing electricity, and running industries. With so much carbon burden, it can be seen that the country would not readily want to accede to emission caps that would help keep the earth's temperature from rising to or above 2 degrees C over pre-industrial levels. Already the earth has warmed by almost 0.80 degrees C since the industrial revolution. The Intergovernmental Panel on Climate Change (IPCC) in its 4th report estimated that to keep temperature rise to 2.0-2.4 degrees C, greenhouse gas emissions must be cut by 50 to 85 per cent relative to 2000 levels by the year 2050. If nothing is done to check the rise in temperature, up to 30 per cent of plant and animal species would be under threat of extinction.

Carbon markets and other mechanisms

The Clean Development Mechanism (CDM) can sometimes be applied to ridiculous extremes. We take for example the marketing of gas flare stopping projects as CDM projects. Through such projects, oil corporations and the Nigerian government hope to claim carbon credits for helping fight climate change. The reality is that gas flaring has been an illegal activity in Nigeria since 1984 when the Nigeria law on Gas Reinjection came into effect. Any reduction or stoppage of flaring is simply a reduction or halting of a criminal activity and brings on no additionality, as the CDM process requires. Any compensation for such an activity flies in the face of reason. Gas flares are the most cynical manifestations of corporate insolence in the face of climate change and environmental health. The flares release greenhouse gases such as carbon dioxide, methane and nitrous and sulphur oxides. Apart from these, the flares release other harmful substances that greatly affect human health.[4]

Environmental Rights Action (ERA), after reviewing the projects for which the oil companies seek carbon credits, concluded that the CDM projects being pursued by the oil companies would only create perverse incentives for the continuation of the obnoxious act of gas flaring. The group noted that in its 2010 Sustainability Report, Shell recorded a 30 per cent increase in gas flaring from their operations in Nigeria compared to their figures in 2009. This happened despite Shell's gas gathering projects and a CDM project at Afam that they commissioned in 2009.

Furthermore, the projects create the need for new oil wells to be exploited

4. http://nnimmo.blogspot.com/2012/11/human-rights-and-multiple-environmental.html (accessed 7 June 2016)

to supply gas to the CDM facilities. ERA noted that all of the registered CDM projects have at least a 10 to 21 year contractual tenure in the first instance to either supply gas or generate electricity. The implication is that these CDM projects constrain the ability of government to make laws or implement laws made for the good governance of the country within this contractual period.

ERA's analysts also see the CDM projects as allowing the Nigerian government and its policy makers to "honestly believe" or to hide behind the illusion that carbon emission can continue since the same amount of pollution can be offset by fossil fuel projects such as those at Afam, Kwale, Oben and Ovade CDM project, or via a carbon sink that would absorb the pollution in future, or as long as they could get some more foreign exchange from them.

Beginning from the market mechanism known as the Clean Development Mechanism (CDM) we are now seeing offspring such as the Reducing Emissions from Deforestation and Forest Degradation (REDD). A number of other mechanisms fall in between, dealing with one carbon off setting mechanism or the other. These market mechanisms provide the basis for private sector investment in projects that appear to have positive possibility of tackling climate change.

The European Union pushed through an empty Green Climate Fund with no money to help countries least responsible for climate change adapt and pursue low-carbon development. The package, however, ensures the possibility that multinational corporations and international financial actors would be able to access any funds that might eventually materialise.

Agrofuels and landgrabs

The phenomenon of land grabbing has risen most as a result of the wrong approach to tackling global warming. While climate change has induced food deficits on account of unusual weather events, the thinking that agrofuels can replace fossil fuels has driven investors to grab lands in the tropics for the cultivation of fuel crops such as sugar cane, jatropha and the like. Up to 200 million hectares of land have been grabbed in this way globally.

It is clear that this phenomenon will not help but worsen the food crisis, land conflicts, the displacement of labour and the poor.

Land grabs create strife and also weaken capacities to mitigate or adapt to the impacts of climate change. Consider cases where top political leaders are directly involved in land allocations or act as middlemen in such transactions. An example is the land crisis that has dogged Uganda. In 2007 there was a massive crisis over plans to convert Mambira forest into a sugarcane plantation. Mid 2012 another land crisis and mass evictions rocked the country with the government threatening to deregister Oxfam and the ULA. Reports have it that Oxfam has had brushes with the government when they insisted that programmes should support the way of life of the people rather than pushing them into patterns that they are ill-equipped to deal with.

Templates for action

Official negotiations have locked in action until possibly 2020. By that time the world may already be in the throes of runaway climate change. Hope is not lost, however. Long before the Arab Spring and the Occupy movement (manifested in many parts of the world and still going on at the time of publication of this book) showed that the regaining of peoples sovereignty over political structures is possible, there was an epochal World People's Conference on Climate Change in Cochabamba, Bolivia, in April 2010.

The Cochabamba conference produced the People's Agreement as a key outcome. It demanded that countries cut their emissions by at least 50 per cent at source without resorting to carbon offsets and other trading schemes. In terms of financing adaptation and mitigation, the Agreement required that developed countries commit six per cent of their GDP for these needs. These plus a reduction in military budgets and committing the savings towards tackling global warming would provide sufficient funds for the so-called Green Climate Fund.

The peoples of the world equally highlighted the necessity of recognising and paying the huge climate debt piled up by the rich polluting nations. Besides making funds available, it will be a step in the direction of decolonizing or democratising the atmosphere of which developed countries have already taken up 80 per cent of the carbon space.

In order to work to restore the natural cycles of Mother Earth, the Agreement demanded the Universal Declaration of the Rights of Mother Earth. This is already being promoted in the UN systems.

Mother Earth or barbarism

Explaining the concept of Mother Earth, Alvaro Garcia Linera, the Vice President of Bolivia, said, The concept of Mother Earth 'is not just a slogan. It means a new way of producing, a new way of relationship with nature and with one another. This relationship is one of equality and not domination, a relationship of dialogue, of giving and receiving. It is not merely a philosophy or folklore'. It is a new ethics, a new way of developing technologies and modes of production. Recalling a statement by Rosa Luxemburg, ssocialism or barbarism', Vice President Linera said that today we could say 'Mother Earth or barbarism.' Affirming that capitalism was the root cause of climate change and many of the ills of the world today, Linera said that the system permits oil companies and the military complex to commit genocide, destroy the environment and reap ever-rising profits at the expense of the blood of the people.

Reinventing governance

We cannot wish away the rapid environmental changes occurring today. There is an urgent call for actions in line with the realisation that planet earth is populated by interdependent beings and cycles. Massive contaminations have already hit as we have already noted. While we cannot exhaust the long list of fronts where the assaults are intensifying, we must mention that critical concerns continue to rise in area of genetic contamination, geo-engineering as well as synthetic biology and the related bio-economy.

It is time for a global rejection of current energy consumption where environmental costs and social liabilities are externalised and rather invest in and build the eco-logic model where ecology, sovereignty and good living define relationships of production and consumption. We need to rebuild our collective environmental and social consciousness, moving away from a system that destroys society and nature through the destruction of knowledge and positive productive forces. We need a system with conscience.

In sum, the struggle is about who defines your narrative of life and living. Who knows best where the shoe pinches – the shoemaker, shoe seller or the person wearing the shoe? Global solidarity is needed and the peoples must regain leadership in this. After decades of inaction to address the pains of the people, leaders now have to be led by the people. Governance structures require reinventing. Inaction simply gives the space for the environmental changes to consolidate as humankind's hangmen.

4

Africa in the vice-grip of the climate crisis

This paper written in 2012 and used for staff training of Environmental Rights Action

Climate impacts on Africa and other vulnerable regions are not matters of speculation. The impacts are real. Failing rains, failing crops, increased desertification and population displacements are some of the manifestations. These impacts are not letting up and will not diminish until something is done. The big question is will something be done?

For Africa, perhaps more than anywhere else, climate impacts are not merely environmental. Climate change impacts diverse realms of life and living. These include poverty, hunger, peace, security, human rights, health and socio-economic aspects.

Globally, the causes of global warming are generally understood. It is also generally agreed that the phenomenon is a global one that requires global action. The big gulf is with regard to what actions must be taken to fight the menace. The blockage to needed for action appears to be erected by the perceived sense of inbuilt resilience by the rich and industrialised nations who feel they can withstand the ravages of global warming over several decades, making it easy for policy makers to delay action while hoping that their grandchildren would help themselves when the tide eventually threatens to swallow them up. Another factor is the strong grip of the fossil fuels sector over policy structures. Through well-orchestrated spin they sell the idea that the continuation of the fossil fuel path is inevitable. They dismiss examples being shown that renewables can and must replace the fossil driven mode of production and consumption. Policy makers who swallow the deception of the fossil fuel sector would do well to pay attention to what analysts have said of the critical need for urgent transition. Any delay is just that, a delay. The problem is that some people must pay the cost. And the most vulnerable, the victims, will bear that cost through their blood, miseries and tears.

In the words of Hermann Scheer:

> If the transition from the nuclear and fossil to renewable energy is only carried out in a piecemeal and gradual manner, then it is highly likely that world civilization will be thrown into a staggering

crisis affecting everyone and everything: dramatic climate change threatens to make entire habitats unfit to live in and to trigger mass misery and the migration of hundreds of millions of people.[1]

The addiction to fossil fuels path is leading to desperations in the oil fields with oil companies moving into more fragile ecosystems, into deeper offshore and extending their claws into the Arctic. These moves are already accompanied by more polluting accidents that compound the environmental crisis. Meanwhile oil companies dither to squeeze the last drop from their oil concessions, the last coal from the mines and engage in dangerous extraction of shale gas as well as tar sands.

There is no disputing that the upsurge in global temperatures is due largely to the amount of carbon released into the atmosphere. Carbon is a basic building block in every living thing, plant or animal. Our soils are loaded with carbon and so is our air and oceans. We take in oxygen and exhale carbon dioxide. Plants do the reverse and we coexist happily supplying each other's carbon dioxide or oxygen needs. The problem is that over the last two centuries humans have dramatically increased the amount of carbon dioxide and other greenhouse gases released into the atmosphere. It is estimated that about 26 billion tones equivalent of CO_2 is released this way yearly. One way to visualise how CO_2 plays a vital role in the climate equation is to see the gas as enveloping the entire earth, occupying a belt around it, so to speak, in the lower atmospheric regions. Now, there are several other gases there, but the essential difference is that this gas, along with other greenhouse gases, allows the energy from the sun through to reach the earth but slows down the escape back into space of the energy reflected from the earth's surface. This build up of energy is the greenhouse effect.

It is not difficult to fully grasp the importance and magnitude of hydrocarbons. Modern urban life is petroleum-based: it depends on it for electrical power and transportation, and releases petroleum in its 300 million tons of waste annually. Modern life is petroleum-based: it depends on machinery, agrochemicals such as the 136.44 million tons of fertilisers plus millions of tons of insecticides, herbicides, fungicides and other chemicals used annually, as well as the transportation of agricultural products. Healthcare and food systems are becoming ever more petroleum-based as food and health sovereignty are increasingly abandoned. In the United States alone, coal is the source of half of all the electrical power generated.

Climate change is fundamentally a systemic crisis that cannot be tackled through palliatives. It is a crisis of a civilisation built on the rapacious destruction of nature through massive consumption and waste of resources. It is the consequence of ruinous production and consumption patterns that are blind to

1. Scheer, Hermann (2012) *The Energy Imperative – 100 Per Cent Renewable Now*, London, Earthscan, p.3

the fact that planet earth is finite and that there must be a point when such a path becomes unsustainable in many ways.

Adaptation and mitigation action taken to tackle climate change impacts are nothing more than palliatives–a term most Nigerians have come to grips with as something done to ease pains inflicted on the peoples by acts of poor governance as well as institutionalised perfidy. These palliatives do not in any way solve the problem; they only help to cope with certain degrees with the crisis.

Systemic challenges, systemic solutions

The global nature of climate change necessitates actions of global proportions and across the earth. Climate change is a crisis of capitalism and its attendant creed of expansion and unlimited economic growth and profits. The current inability to confront this crisis is due mainly to the vice grip on the global systems by the powerhouses of imperialism. The United Nations Framework Convention on Climate Change (UNFCCC) was set up to provide guideposts and to urge nations to act together in everyone's common interest for the survival of the planet as we know it. The conferences of parties (COP) to the convention have quickly turned into talk shops for nations to display their power, arm-twist the poor and evade action. This became most obvious at COP15 in Copenhagen and became entrenched at COP16 in Cancún. What happened at COP17 in Durban must take the medal as a conference whose critical achievement was the blatant postponement of action while the earth burns.

Climate justice activists saw the Durban debacle as one where ordinary people were unabashedly let down by governments. Leading the betrayal were the governments of the USA, Japan, Australia, Canada and other developed nations who reneged on their promises, weakened the rules on climate action and strengthened those rules that allow their corporations to profit from the climate crisis.

At the close of the Durban meeting, Sarah-Jane Clifton of Friends of the Earth International (FoEI) explained for example that the Kyoto Protocol, the only legally binding framework for emissions reductions, survived in name only, and

> The ambition for those emissions cuts remains terrifyingly low. The Green Climate Fund has no money and the plans to expand destructive carbon trading move ahead. Meanwhile, millions across the developing world already face devastating climate impacts, and the world catapults headlong towards climate catastrophe. It is clear in whose interests this deal has been advanced, and it isn't the 99 percent of people around the world.The noise of corporate polluters has drowned out the voices of ordinary people in the ears of our leaders.[2]

2. Friends of the Earth International (2011), 'Reaction to Durban climate talks', 11 December,

While Mohamed Adow of Christian Aid saw the Durban outcome as "… a compromise which saves the climate talks but endangers people living in poverty," the UNFCCC secretariat saw COP17 as a great success and indeed called it

> … a landmark historical COP – not just the longest but also the decisions here have really marked a completely new trajectory for the climate regime. It has guaranteed a second commitment period but it has also laid the path for a broader regime applicable to all in a legal way and provided mechanisms for developing countries to address their needs of mitigation and adaptation.

The response of analysts such as Pablo Solón, former lead negotiator for the Plurinational State of Bolivia, was clear and to the point: 'It is false to say that a second commitment period of the Kyoto Protocol has been adopted in Durban. The actual decision has merely been postponed to the next COP, with no commitments for emission reductions from rich countries. This means that the Kyoto Protocol will be on life support until it is replaced by a new agreement that will be even weaker.'[3]

The descent into non-binding pledge-and-review system places the world on the way to warming by as much as 4 to 7 degrees C above preindustrial levels. If that happens, Africa will be cooked because the continent experiences 50 per cent more than average global temperature levels. The reluctance of rich and highly polluting nations to take real actions on climate change has rightly been described as a form of apartheid. This is apartheid against Mother Earth and the species that she bears.

While the people see the outcomes of COPs 15, 16 and 17 as off the mark, officials insist that sitting in the driving seat is not the same as being in the passenger's seat. They say that reaching those conclusions in a conference of so many governments ought to be applauded. The views of key players at the conference are revealing. Let us look at samples of these:

For Maite Nkoana-Mashbane, the president of the Durban conference: 'We have taken crucial steps forward for the common good and the global citizenry today. I believe that what we have achieved in Durban will play a central role in saving tomorrow, today.'[4] The executive secretary of the UN

http://www.foei.org/press/archive-by-year/press-2011/reaction-to-durban-climate-talks (accessed 28 May 2016)

3. Focus on the Global South (2011), 'CJN: COP17 succumbs to Climate Apartheid. Antidote is Chochabamba Peoples' Agreement', http://focusweb.org/content/cjn-cop17-succumbs-climate-apartheid-antidote-cochabamba-peoples-agreement (accessed 28 May 2016)

4. Conway-Smith, Erin (2011)'China Says Rich Countries don't care Enough About Climate Change To Do Anything', *Business Insider Australia*, (12 December), http://www.businessinsider.com.au/china-says-rich-countries-dont-care-enough-about-climate-change-to-do-anything-2011-12 (accessed 28 May 2016)

Framework Convention on Climate Change (UNFCCC), Christiana Figueres was no less upbeat: 'I salute the countries who made this agreement. They have all laid aside some cherished objectives of their own to meet a common purpose — a long-term solution to climate change. The South African presidency steered through a long and intense conference to a historic agreement that has met all major issues.[5]

The water and environment affairs minister of South Africa, Edna Molewa, waxed superlative in her summation of the COP: 'After a year of intensive negotiation, the final outcome of Durban is historic and precedent-setting, ranking with the 1997 conference where the Kyoto Protocol was adopted. In the dying hours of this watershed conference, we were able to agree on a comprehensive deal.'[6]

The successes of the Durban conference alluded to:

1. Agreement to set up a Green Climate Fund
2. Agreement to negotiate a binding emissions reduction by 2015 and bring it into force by 2020 (an 'agreement to negotiate' is a quaint form of agreement)
3. Not burying the Kyoto Protocol (but merely leaving it on life support)
4. Jettisoning of equity aspect of the Kyoto Protocol (by erasing the common but differentiated responsibilities provisions)
5. Agriculture got mentioned in the outcomes (for carbon credit purposes opposed by farmers and mass movements)

Would you trust a COP?

In *To Cook A Continent*, I noted:

The question that confronts humanity is how to tackle this carbon in the atmosphere. What should people do? Do we stop releasing carbon into the atmosphere? What would that entail? Are there acceptable levels that we can keep to so as not to reach the point of no return with a runaway climate change? These questions have engaged the imaginations of many and a variety of solutions have been offered.

One interesting proposed solution that has come up is that of carbon sequestration. In this scenario you could keep up with business as usual, releasing as much carbon as you please, only ensuring that the carbon is captured and stored or sequestered. Technologists have been at work on carbon sequestration technologies with some already undergoing tests, but

5. United Nations (2011) 'Durban conference delivers breakthrough in international community' response to climate change' https://unfccc.int/files/press/press_releases_advisories/application/pdf/pr20111112cop17final.pdf (accessed 28 May 2016)

6. Bond, Patrick, 'Durban's conference of polluters, market failure and critic failure', ephemera, http://www.ephemerajournal.org/contribution/durban's-conference-polluters-market-failure-and-critic-failure (accessed 28 May 2016)

with none likely to be ready for practical use until about 2020. Meanwhile carbon offset is gaining ground in some quarters with the proposition that one could pollute in part of the world, and then have the pollution offset in another part of the world. For example, a company in Europe could keep on stoking the atmosphere with carbon, but plant a tree plantation somewhere in Africa and since the trees absorb carbon, the company can feel satisfied that they are carbon neutral. It is a convenient fictional scenario. The company cleans up its conscience; those setting up the plantation get paid. Communities whose lands are taken up to set up the plantations may get jobs as plantation hands and perhaps receive some form of compensation for lost farmlands.[7]

The COP17 outcome seeks to include developed and developing nations in set emissions reduction targets. But it would only negotiate such a new legally binding agreement in 2015. And whatever this agreement turns out to be, it will only take effect after 2020. This means that from now until 2020, nations would cut emissions based on their own national pledges, which are voluntary and not legally binding.[8]

Thus the Durban conference eliminated any sense of equity and fairness predicated on the common but differentiated responsibility in the efforts to tackle climate change.This was done despite the fact that the rich industrialised countries are responsible for three quarters of all emissions historically whilst hosting only 15 percent of the world's population.

The Durban outcome showed no sense of urgent action in the face of the planetary emergency— just a promise to start a whole new round of negotiations on a whole new treaty, delaying action until 2020— by which time we may already have irreversible crisis on our hands.

The entire foundation of the climate regime was pulled apart, and those to blame for this destruction are the governments of United States, Canada, Japan, Australia and the other rich industrialised nations who worked hard to avoid their legal and moral obligation to make deep and urgent emissions reductions to help the world avoid the onset of runaway climate change.

By forcing the so-called Durban Platform through the UN process, the United States and the EU led the charge in shifting the burden of tackling climate change squarely onto the shoulders of countries in Africa and the global South.

While countries dither about what can be done globally to avoid a 2 degrees Celsius rise in global temperatures others have learned from the Katrina floods that showed the soft underbelly of the USA that no one is immune to what is coming in the future. Thus, Germany is spending 600 million Euros on one new sea wall for Hamburg while The Netherlands plans to spend 2.2 billion

7. Bassey, Nnimmo (2012) *To Cook A Continent – Destructive Extraction and the Climate Crisis in Africa*, Oxford, Pambazuka Press, pp.105-106

8. This emissions pledges and non-committal choreography was confirmed at COP20 and COP21

Euros on dykes between now and 2015. These measures may be inadequate considering they are indicators of what rich nations are capable of doing in a bid to survive the coming floods. The question is, what will happen to the poor? What is the response to the cries of the Small Island States such as Tuvalu whose citizens are already becoming climate refugees?

REDD

Reducing Emissions from Deforestation and Forest Degradation (REDD) is one mechanism that allows polluters to keep polluting, take no action to reduce emissions but rather buy carbon credits that enable them to carry on with business as usual. Many observers including the Durban Group for Climate Justice exposed this loophole and denounced REDD as a false solution. In 2009 at Copenhagen the Durban Group stated:

> Like Clean Development Mechanism credits, they exacerbate climate change by giving industrialised countries and companies incentives to delay undertaking the sweeping structural change away from fossil fuel-dependent systems of production, consumption, transportation that the climate problem demands. They waste years of time that the world doesn't have. Worse, conserving forests can never be climatically equivalent to keeping fossil fuels in the ground, since carbon dioxide emitted from burning fossil fuels adds to the overall burden of carbon perpetually circulating among the atmosphere, vegetation, soils and oceans, whereas carbon dioxide from deforestation does not. This inequivalence, among many other complexities, makes REDD carbon accounting, impossible, allowing carbon traders to inflate the value of REDD carbon credits with impunity and further increasing the use of fossil fuels.[9]

REDD does not seek to halt deforestation as it rather focuses on what the promoters believe is the carbon stock in the trees. In this process, forest communities suffer evictions and exclusion from their territories and forest resources. Moreover, the United Nations sees monoculture plantations as forests, making it attractive for carbon speculators to convert forests into plantations in efforts to disconnect indigenous communities from ever claiming titles over their patrimony. REDD is a major driver of land grabbing that is impoverishing already poor communities and deepening the food crisis by converting arable lands for cash cropping and related projects aimed at the export markets.

While communities stand to lose, big companies are raring to go in their carbon cash dash. Big polluters investing in REDD include Shell, Rio Tinto and Chevron-Texaco who see this as an avenue for buying their way out of reducing their greenhouse emissions at source by supposedly conserving forests.

---

9. See 'The Global Alliance of Indigenous Peoples and Local Communities against REDD', (07 December 2011), at http://www.iucn.org/about/union/commissions/ceesp/?8786/The-Global-Alliance-of-Indigenous-Peoples-and-Local-Communities-against-REDD (accessed 28 May 2016)

The oil giant Shell (well known for its role in the environmental destruction of Nigeria's Niger Delta), Gasprom and the Clinton Foundation are funding the landmark REDD Rimba Raya project in East Kalimantan in Indonesia.[10]

Agrofuels – the new fossil fuels?

In 2006 the former president of Senegal proposed the creation of what he termed a 'green' OPEC expected to be the Pan-African Non-Petroleum Producers Association (PANPP). President Abdoulaye Wade, in an article titled Africa Over the Barrel, envisioned the body as one that would aspire to 'become leaders in the field of biofuels and alternative energy strategies, following in Brazil's footsteps.'

The notion of a green OPEC suggests a propensity for commodity cartelisation as well as a fixation on a model dependent on fossil fuels. With fifteen countries already on board and meetings already held, the PANPP dream may wax strong. At the founding meeting, twenty-five countries participated, with fifteen endorsing the founding charter. The countries included: Benin Republic, Burkina Faso, Democratic Republic of the Congo, Gambia, Ghana, Guinea, Guinea-Bissau, Madagascar, Mali, Morocco, Niger, Senegal, Sierra Leone, Togo and Zambia.

The concerns are whether its activities will help or worsen the food crisis and land conflicts on the African Continent.

We were very disappointed that the Copenhagen conference attempted to impose the adoption of an accord on the COP following an undemocratic process.The Copenhagen Accord attempted to eliminate the Kyoto Protocol and ignored concerns already well shown by science and equity to be critical for the survival of the earth and her peoples. The 'Accord' does this mainly through the negation of the principle of aggregate emissions reduction target for rich nations. By promoting voluntary emissions targets, the Accord has so far pointed at a possibility of a temperature increase of over 4 degrees Celsius – a clear notice to Africa that she should prepare to be incinerated. Comparable extreme fate awaits small island and Arctic states if that scenario is allowed to stand.

The Peoples Agreement, a key outcome of the Cochabamba conference, demands that countries cut their emissions by at least 50 percent on 1990 levels at source in the second commitment period of the Kyoto Protocol (2013-2017), without recourse to offsets and other carbon trading schemes. These market mechanisms such as those in CDM and REDD sell false solutions to speculators who are happy to do nothing while exposing the earth to grave danger through unabated greenhouse gas emissions.

10. Lang, Chris (8 September 2010). Shell REDD project slammed by indigenous Environmental Network and Friends of the Earth Nigeria. www.redd-monitor.org/2010/09/08/indigenous-environmental -network-and-friends-of-the-earth-nigeria-denounce-shell-redd-project/ (accessed 28 May 2016)

In terms of finance, the Peoples Agreement demands that developed countries commit 6 percent of their GDP to finance adaptation and mitigation needs. The financial suggestions of the Copenhagen Accord are a drop in the ocean compared to what is needed to secure vulnerable peoples and nations.

The peoples of the world also affirmed that there is a climate debt that must be recognised and paid. The payment is not all about finance but principally about decolonizing the atmospheric space and redistributing the meagre space left. Developed countries already occupy 80 percent of the space.

The climate debt is also about taking actions needed to restore the natural cycles of Mother Earth and one clear way of achieving this will be through the proclamation of a Universal Declaration on the Rights of Mother Earth with clear obligations on humans.

The Peoples Agreement recognises that the causes of climate change are systemic and that systemic changes are needed to tackle them. On this note, the model of civilisation that is hinged on uncontrolled development can only compound the crisis. The world needs to move towards living well and not continue on the path of domination of others and of conspicuous and wasteful consumption.

An area glossed over in the UNFCCC negotiations is the role of industrial agriculture in climate change. The Peoples Conference debated this key sector and reached the agreement that the way to a sustainable future is through the enthronement of food sovereignty based on agro-ecological and peasant based agricultural systems.Water is a public trust and an inalienable human right and the continued global destruction of the world's water and its commodification was clearly rejected in the Peoples Agreement.

In all, the Peoples Agreement recognises that real strategies to tackle climate change must be based on the principles of equity and justice in dealing with the structural causes. Without climate justice it will clearly be impossible to achieve the much talked about Millennium Development Goals (MDGs).

Cochabamba resonated with calls for urgently securing the rights of Mother Earth as a means of reconfiguring our relationship with the earth and with each other— in a way that respects the past, today and the future. All these will be a pipe dream unless peoples' sovereignty is supported, restored and built across the world. Cochabamba was a turning point in the march to transform our world from the paths of conflicts, competition, exploitation and domination unto a path of solidarity and dignity.

As representatives of peoples from across the world we urge the United Nations system to embrace the contributions captured by the Peoples Agreement and other outcomes of the Cochabamba World peoples Conference on Climate Change and to use them as the key to untangle the gridlock that is keeping the world on the path to even more climate chaos.

In a presentation to the UN Secretary General, Meena Raman asserted that 'The Peoples' Agreement from Cochabamba has called for a global referendum or popular consultation on climate change, should the process of future climate meetings not deliver the aspirations of the Peoples' Agreement.'[11] She noted that a key outcome at the next COP must be an amendment to the Kyoto Protocol for the second commitment period from 2013-2017 for developed countries to agree to domestic emissions cuts of at least 50 percent from 1990 levels excluding carbon markets or other offset mechanisms. She stressed that the Kyoto model is superior because it mandates an aggregate reduction target for Annex I parties, it mandates each Annex I country have an adequate target which is also comparable with one another, and that these are legally binding. Under the Copenhagen Accord, there is no aggregate target, and each country only pledges what it wants even if it is not adequate. Under the Accord the pledges as you know are grossly inadequate, adding to only a 11–18 percent cut compared to the 40 percent cut needed. Raman concluded that the continuation of the Kyoto Protocol was imperative and noted that developed countries, through the Copenhagen Accord, were attempting to 'kill' the KP.

Conclusion

The negotiations at the UNFCCC COPs continue to drag away from real solutions and instead move deeper into offset regimes, depending on market forces that have clearly driven the world into financial, economic and political crises. Larry Lohmann[12] reminds us that when Sir Nicholas Stern, climate change advisor to Tony Blair's government, famously said in 2007 that global warming was 'the greatest market failure the world has ever seen', the implication was that, given the proper price signals, addressing it could be a market success.

Lohmann adds that with that mindset, predictably, Sir Stern became a climate entrepreneur, serving also as advisor to IDEAcarbon, a company whose ambition is to provide 'ratings, research, and strategic advice' on carbon commodities and finance to 'buyers, sellers and hedgers'.

We need to evaluate the creed that presents the market as holding the solution to every problem and the key to human progress and happiness in the face of the crises it has spawned. Humans cannot trade their way out of the impending climate catastrophe.

The fossil fuel sector needs to be reined in. In particular, the oil sector that most entrenches the production and consumption patterns that aggravate global warming is very closely interwoven with the military industrial complex. The intertwining is so deep that a map of flash-points in the world would almost equal one of crude oil and gas fields. Thus, when the UNFCCC creates

11. See the Submission by the Plurinational State of Bolivia to the Ad-Hoc Working Group on Long-Term Cooperative Action at http://unfccc.int/files/meetings/ad_hoc_working_groups/lca/application/pdf/bolivia_awglca10.pdf (accessed 7 June 2016)

12. Larry Lohmann. 28 October 2011. *The Endless Algebra of Climate Markets.* http://www.thecornerhouse.org.uk/sites/thecornerhouse.org.uk/files/LohmannCNSarticle.pdf (accessed 7 June 2016)

a climate account with no cash in it rich countries are busy investing billions of dollars in destructive wars and selling weapons to struggling economies with no need of standing armies.

Less than ten per cent of the budgets lavished on wars by rich nations would more than take care of helping vulnerable nations build resilience as well as adapt to the unfortunate situation they are driven into. The same can be said of the funding of the MDGs.

It is time for world leaders to listen to their people. Climate change does not divide the peoples of the world. It unites them in more sense than one. It offers a clear opportunity for solidarity actions to stem the tide and show the present generation of humans cares about the future.

The current trend wherein development is measured in terms of pollution must be halted. Levels of carbon emission should be a measure of irresponsibility and not one of well-being. The more polluting a country is, the more irresponsible it should be seen to be and such countries should be held accountable for their impacts on the planet.

The real solution to climate change is the cutting of emissions at source rather than wasting resources on untested technologies such as those of carbon capture and storage and geo-engineering. This is the time for the world to quickly move away from the fossil fuels driven civilisation. The equation of energy security to national security will continue to lead some nations into military adventures that, apart from being destructive, themselves consume huge fossil fuels and compound the problems of climate change. The neoliberal system permits the World Bank to parade itself as a climate bank while it keeps funding dirty energy projects such as the Eskom coal plant in South Africa and a number of other fossil fuels projects elsewhere. It is time for the overturning of corporate power and halting of its erosion of peoples' sovereignty.

The world needs urgent actions to ensure rapid transiting from the current fossil fuels driven path and build transformation solutions on a renewable and sustainable path. The key nodes of this agenda will include:

1. Reclaiming peoples control over their resources
2. Building progressive people-oriented governments and power structures and shifting away from destructive modes of relations
3. Redirecting military budgets for climate change mitigation and adaptation
4. Legislation – such as the Rights of Mother Earth
5. Recognition of ecocide as an international crime and the commencement of the prosecution of eco-criminals akin to what happens at the current International Court of Justice, but not allowing any nation or corporation to opt out.
6. Leaving fossil fuels in the soil
7. Having binding agreements for emissions reduction, and jettisoning carbon offsetting as a means of tackling global warming.

5

Where are the 50-year-old trees?

This article was written in September 2010 for 234NEXT and was also published in October 2010 by TLAXCALA[1]

Trees are the lungs of the earth; it can be assumed then that Nigeria, which displays severe cases of deforestation, is literally gasping for breath for lack of oxygen. Nigeria's rainforests have been depleted to less than 10 percent of their size 50 years ago.

What is left of our forests are under threat, and many areas are degraded and converted for other uses. This phenomenon is not restricted to Nigeria. Overall, the United Nations surmises that 13 million hectares of forested land have been converted every year over the past 10 years and most of this is said to be for agricultural purposes.

In areas like the Amazonia, most of the conversion has been for monoculture farms where crops like soy are cultivated for animal feed and other industrial uses. In South East Asia, the threat has been from forest conversion into oil palm plantations for ultimate production of fuels for machines. Indeed, the land uptake and the highly invasive oil palm trees, make this writer wince whenever Latino friends insist on calling the tree 'palm Africano' or 'African palm'. A crop, whose origin is Africa, is being introduced in environments where the locals view its arrival with anxiety and at times anger.

In Nigeria, our forests are threatened by logging for export and conversion into plantations. It is said by some that we export logs and import back into Nigeria as floorboards, furniture, or even toothpicks. As The Economist noted in a recent issue, 'clearing forests may enrich those who are doing it, but over the long run it impoverishes the planet as a whole.'

There was a notorious forest gobbler who was well known for logging outside its area of concession in the Omo Forest. Whenever confronted, the operators of this company would plead that they simply got lost in the forest. Interestingly, they always 'innocently' strayed into areas where the bigger trees were.

The company's creative way of avoiding responsibility perhaps reached its crescendo in the Cross River forest, after they found the Omo Forest too

---

1. http://www.tlaxcala-int.org/article.asp?reference=1946 (accessed 7 June 2016)

hot to stay in. When community folks in Cross River began to complain about this company's activities, they recruited a bunch of community people who supported the company and even staged a demonstration during which they carried signs saying 'we want factories, not monkeys.' This company was eventually kicked out of the forest by the Cross River State government.

How about those converting forests into rubber plantations? Chunks of the Okomu Forest as well as the nearby Iguobazuwa forest have been at the mercy of the Nigerian subsidiaries of a French multinational tyre company, as they convert the forests into rubber plantations. Some people push the idea that plantations are the same as forests and that they provide the same services. Truth is that plantations are not anything near to being forests; even if the Food and Agriculture Organisation (FAO) of the United Nations says otherwise.[2] At a very fundamental level, a tree cannot make a forest, even if you plant a million stands of that tree. The rise of monoculture plantations has seen huge use of agrochemicals, including pesticides that eat at the very heart of forest ecology.[3]

Plantations do not have the sort of ground cover and undergrowth that forests have, and cannot absorb as much carbon as forests do. Plantations do not provide vegetable, medicinal, and other support to communities. Neither do they provide the essential service of protecting watersheds as forests do. We need not mention that bush meat is not found in plantations.

Time to ask questions

As Nigeria celebrates 50 years of political independence, it is apt to ask if we can easily find many 50 years old trees standing anywhere within our borders. Some would likely be found in community-managed forests; especially those designated as sacred or even evil forests.

The massive loss of our forest cover should give us serious concern as we clink glasses in celebrations that many Nigerians are unable to identify with because of the sad and alienating records we have chalked up in various sectors, including the environmental area.

As we celebrate 50 years of political independence, we may as well look back at the many oil spills that have flooded the Niger Delta over the same period of time. At 50, we should ask why half the population of Jigawa State should be displaced by flooding and why water levels in dams in the northern parts of our country are not properly managed.

At 50 years, we should perhaps place garlands on the necks of multinational oil companies who had flared gas in the oil fields routinely over the same period without care that we are daily choked and killed by the toxic cocktail that they spew into the environment.

2. See 'Forest Plantations', for example, at http://www.fao.org/docrep/004/y1997e/y1997e08.htm (accessed 7 June 2016)

3. See WRM.2009. Exposing the Lies about Monocultural Tree Plantations. http://wrm.org.uy/oldsite/plantations/21_set/Exposing_lies.pdf (accessed 7 June 2016)

As we celebrate 50 years of political independence, we should track how we have fared in all areas of human development. It is a good time to pause and ask if 50 years is not enough for Nigeria and other celebrating African nations to pause and ask when they will have true socio-economic independence.

As we celebrate 50 years of independence, it is a good time to ask what are the ecological agendas of those who wish to contest for the presidency and other offices in Nigeria. The environment is our life, and we cannot afford to have *Honourables* and *Excellencies* who do not care about ecology beyond how to guzzle ecological funds and use them for electioneering campaigns.

It should be a good time to look for a 50-year-old tree anywhere it can be found, sit on its roots, and think. If we cannot find such tress, then we may stand before any sapling we find and promise to let it stand for at least 50 years.

6

To stop the Sahara

This article was published in 16 September 2010 in 234NEXT online newspaper, Lagos[1]
The desert is not an organism that spreads its tentacles to swallow up land and objects in its path. This is the image we generally have of the desert when we speak of the advance or spreading of the Sahara desert. The process by which an area becomes a desert is known as desertification. This process can happen in an area that is not contiguous to an existing desert. In other words, an area that is not close to an existing desert can become one through the process of desertification.

Desertification is one of the key environmental challenges facing Nigeria and indeed all of sub-Saharan Africa. It has been estimated that the desert area is increasing at the rate of more than half a kilometre every year[2], and that about 35 million Nigerians are directly affected by this menace. Eleven states in Nigeria, sometimes called the frontline states, are under threat of desertification. These states include Bauchi, Borno, Jigawa, Kano, Katsina, Kebbi, Sokoto, Yobe, Zamfara, and parts of Gombe, and Jigawa.

Obviously, the problem of desertification has global implications and that is why we have the United Nations Convention to Combat Desertification (UNCCD) that was launched in 1994 and became operational three years later. Although a majority of nations have ratified the convention, only handfuls have undertaken programmes towards the attainment of its objectives. For example, Nigeria took steps in this direction in 2001 when the National Action Programme (NAP) on desertification was launched. Nigeria also has a number of other initiatives: National Grains Reserve Programme; Drought Management Policy; Desert-to-Food Policy; and the National Desertification Policy. A question that comes to mind is how well are these policies and programmes being run?

---

1. http://234next.com/csp/cms/sites/next/opinion/5619066-184/oil_politics_to_stop_the_sahara.csp (website is no longer available)

2. UN.1997. Nigeria Country Profile. http://www.un.org/esa/earthsummit/nigeriac.htmaccessed 29 May 2016[2] Nick Ashton-Jones. 1999. *The Human Ecosystem of the Niger Delta – An ERA Handbook.* ERA: Benin City.

Generally, Nigeria's most visible actions to fight desertification have largely been about organising yearly tree planting exercises. The planting of trees is a good step and should be encouraged. The sad fact is that tree planting alone is not enough to stop desertification from taking place.

We must look at the factors that encourage desertification. Without much investigation, it is obvious that global warming has a major impact on this process. Other factors include bush burning, inappropriate grazing, and poor irrigation systems.

Going back to the contribution of global warming to this phenomenon informs us that Nigeria faces a peculiar risk of being swallowed up by two migrating forces – water from rising sea levels, and sand from increased desertification. Some of the impacts of global warming include droughts as well as freak rains. Even with the unpredictable rainfall patterns, northern Nigeria still has much less rain than the south. The lack of rain directly encourages desertification, as already stated. Along the coastal fringe of Nigeria, the big challenge is that of sea level rise. When we talk of sea level rise, it is vital to keep in mind that the Niger Delta, for example, is a naturally subsiding environment. With this in mind, it has been estimated that if a net sea level rise of one meter occurs, up to 100 kilometres from the Atlantic shore will go under water.

The implication of this is that our economy, food systems, security, and livelihoods are severely threatened by the impacts of global warming. This is an issue that should be taken as a serious matter of emergency and should be tackled holistically, and not by seasonal, episodic responses.

One of the most visible issues relating to desertification in Nigeria is the shrinkage of Lake Chad.[3] This lake has shrunk by more than 90 percent since the past 50 years and we can suggest here that the fortunes of Lake Chad should be taken as a measure of progress made by our country since political independence was attained 50 years ago. Lake Chad has shrunk from an area of 25,000 square kilometres to a paltry 1,500 square kilometres. Experts believe that at its present rate of shrinkage, Lake Chad may become dry land within the next 20 years.

The drying up of Lake Chad and other water bodies is not a mere geographic reality. It means increasing loss of livelihood, increasing water scarcity, and a veritable pusher of poverty. Moreover, the drying of the lake and persisting desertification portend staggering prospects for our nation. With the understanding that the Sahara is not 'marching' but that desertification occurs autonomously, we can take steps to halt this process that allows sands to swallow our land.

The implication of the drying of Lake Chad is already obvious in the displacement of fishermen, pastoralists, and others who depend on its water. The future is bleak, unless the root causes of this phenomenon are tackled.

3. UNEP. 2008. 'Chad: almost gone' in *Vital Water Graphics*. http://www.unep.org/dewa/vitalwater/article116.htmlaccessed 29 May 2016

It may even be the case that the recurring land crises in the middle belt can be traced to the environment displacement of populations in these areas and the religious colouration may well be convenient cover for perpetrators of intolerance.

We repeat here that mere trees cannot stop desertification. Indeed, the trees we plant require enormous amounts of water to thrive, although we can use drought resistant species. You can imagine how much land cover can be achieved by simply planting drought resistant shrubs!

We will end this piece by returning to the issue of global warming as a major contributor to desertification. Reflect on the fact that global warming is caused by the release of greenhouse gases into the atmosphere. There is one industrial complex that releases massive amounts of greenhouse gases into the atmosphere on a daily basis. We are talking about the gas flares in the oil fields of the Niger Delta. If Nigeria is serious about tackling the issue of desertification, one of the immediate first steps is to stop the activity of gas flaring that have been illegal since 1984.

If we continue to stoke the atmosphere with greenhouse gasses through gas flaring in the South and keep planting a tree belt with the intention of halting desertification in the North, we are clearly wasting salt on porcupine intestine. It will remain a bitter tale.

7

Of floods, dams and the damned

This article was written after the 2012 massive flood in Nigeria that left 300 dead and about two million Nigerians displaced. It was published on 12 October 2012 by Sahara Reporters[1]

The flooding that has ravaged parts of Nigeria in recent months has been characterised as a natural disaster. We do not agree with this assessment. Our summation is that the disaster is manmade and must be recognised as such.

The floods are primarily a result of very poor management of the dams in both Nigeria and Cameroon. If we agree with this position, it should also be agreed that the prescription of more dams would mostly be to open up space for enterprises whose business is the building of dams or the production of materials, like cement, used in building dams. More dams will simply mean more trouble. Dams are not the solution to the flooding we are seeing in Nigeria today. They caused it.

How do we reach this conclusion? The dams that are most culpable in this disaster are the Lagdo dam in Northern Cameroon and the Kainji dam in Nigeria. Although there is no definitive information about when water was released from Kainji and possibly other dams in Nigeria, it is believed that some of the floodwaters have come from these sources.

The warning about the impending release of water from the Cameroonian authorities may have come rather late, but it at least offered some people a chance to relocate or take some of their properties out of the path of destruction.On the Nigerian side, there was no warning and all the blame keeps being studiously heaped on the Lagdo Dam. However, it is clear that the water from the Lagdo Dam would first impact River Benue and then the water would join up with the River Niger at the confluence town of Lokoja.The flooding reported in Kwara State, for example, cannot be attributed to the release of water from the Lagdo Dam. We do not need to deploy some sleuths to identity the culprit / collaborator: Nigerian dams.

One big failure of the relevant agencies of the Nigerian government in handling this disaster, which the president says has displaced 25 percent of the

---

1. http://saharareporters.com/article/floods-dams-and-damned-nnimmo-bassey (accessed 29 May 2016)

Nigerian population, is that they failed to warn downstream communities that the floodwaters were headed to the Atlantic Ocean and would sweep over everything in their path. It is truly mind-boggling that this could not have been known. It is unacceptable that the flooding of Edo, Delta, Bayelsa, Rivers and other states could not have been foreseen and the people warned before the deluge swept in.

The River Niger runs from Guinea, through Mali, Niger and Nigeria before discharging its waters into the Atlantic Ocean.The Kainji dam, built in 1968, has a capacity of 15 billion cubic meters covering an area of 1270 sq kilometres. Much of its waters arrive at the dam in July and in December each year. As with other dams, Kainji dam has inbuilt discharge sluices. In addition, the dam was built with some draining channels to help regulate the volume of water preventing it reaching disastrous levels.

According to a source quoted by Sunday Trust, 'the excess water discharged from this dam account for over 80 per cent of the flood being experienced today at the lower River Niger. The blockage of the natural channel is making the situation critical. There is nothing we can do to control the situation. In fact, should the dam receive any additional water within this rainy season, worse disaster will be recorded. The pressure on the dam is at its peak now. Anything could happen.'[2]

The report further notes that the natural drainage channels have been blocked for five years now and no action has been taken by the Ministry of Power to reopen them. If this is true, we can see that this is an additional cause of the flooding being experienced. The drainage provisions in the dam are meant to help control the level of water in the dam during the peak inflow periods. Where the channels are not in service the waters simply keep accumulating and the end result could be a collapse of the dam.

What must be done to avert further disasters and to avoid compounding the current one? Besides all possible lines of action, there is the primary need for government to recognise climate change as an urgent justice/security issue.What we are witnessing now is a foretaste of what would happen when unusual rains and other weather events kick in more forcefully.

Multi-agencies and stakeholders' actions are needed to curtail or avoid the sort of impacts of flooding we are seeing. While the Ministry of Works should ensure the building of resilient shoreline infrastructure, for example, it is essential for the Ministry of Environment to urgently facilitate the setting up of climate crises committees in coastal communities. These committees would harness community resources towards adaptive actions and also provide guidelines as to where buildings and sewage infrastructure may be constructed.

2. Theophilus Abbah, et al. October 7, 2012. *Danger Looms at Kainji Dam*. Sunday Trust newspaper, Abuja. http://www.dailytrust.com.ng/sunday/index.php/top-stories/11859-danger-looms-at-kainji-damaccessed 29 May 2016

These planning regulation duties would help inculcate in the people a sense of disaster preparedness while building more resilient communities.

A cursory look at many buildings in the flood prone areas shows that the ground floor level of the buildings is either at the same level as the adjoining roads or even below them. Such buildings are extremely flood prone in normal circumstances and stand no chance at all when the water levels rise to the rooftop levels we now see. The method and materials used in building construction also give an indication as to what the owners of such buildings would expect to see when the floodwaters ebb.

The issue of elevation of building floors also pertains to the elevation of roads. Large sections of the new lane of the East West Highway are far below the level of the existing lane. Thus as the flood waters arrived, the sand filled but the yet-to-be-paved areas were quickly washed away.

We recommend that all dams in and around the nation must be reviewed for structural integrity as well as for adequacy of maintenance routines. If the floods are caused by inadequacy of the number of dams, then the new dams should be upstream of the current ones. That would conceivably place those new dams outside the borders of Nigeria. Thus new dams are not the solution to this deluge.

The engineering corps of the Nigerian military should be drafted into emergency restoration of damaged infrastructure. They should also be empowered to urgently commence preparations for post flood restoration interventions. The Nigerian architects, engineers and builders should step in at this time to provide voluntary services just as the medical professionals are doing.

In conclusion we reiterate that this flooding is a man-made disaster and creating more incentives for future disasters is not acceptable. We also repeat that the management of the dams is the major culprit. In addition, climate change is a real partner. Both are man-made, cannot be taken as so-called acts of God and must be appropriately addressed. Otherwise we would literally make our people the damned.

8

Flaring gas: Profiting from illegalities in Nigeria

This article was written in April 2012 and a version of it appeared in MakingIt
Magazine on 15 November 2012[1]

While the world dithers on actions to curb the release of more greenhouse
gases into the atmosphere, oil companies in Nigeria are busy pumping the gases
into the atmosphere through gas flaring and they are reaping huge profits as
they do so.[2]

Gas flaring in Nigeria was confirmed as being unconstitutional and an
abuse of human rights by a High Court in November 2005. These have been
the complaints of communities living in the oil fields of Nigeria over the past
five decades. Indeed, the act of routine gas flaring was outlawed in Nigeria
in 1984 when a law on gas reinjection came into effect. From that date, oil
companies were disallowed from engaging in routine gas flaring and could only
flare on receipt of a permit from the responsible minister. They would also pay
a fine. However, these have not proved a sufficient deterrent.

In 1984 the fine was a mere US$0.003 (0.3 cents) per million cubic feet. It
increased in 1988 to US$0.07 per million cubic feet, and to US$3.50 for every
1,000 standard cubic feet of gas flared in January 2008.

Since the restoration of democratic politics in Nigeria in 1999 many
terminal dates for the halting of gas flaring have been set by successive
administrations with none being honoured. Unfulfilled deadlines include end
of 2007, 2008 and 2010. The Nigerian National Assembly also attempted
proposing a new deadline of 2012 when the Senate passed a bill that effectively
criminalised and raised the levels of punishment and fines in an attempt to make
the fine equal to the commercial value of the gas being flared. That bill could
not see the light of the day due partly to heavy industry pressure according to
observers.

Claims by oil companies that they are working to reduce gas flaring are

---

1. See at http://www.makingitmagazine.net/?p=5890 (accessed 29 May 2017)

2. The flaring of natural gas associated with crude oil extraction has been illegal in Nigeria since 1984.
The obnoxious act continues unabated. In fact, Nigeria is second only to Russia in gas flaring with more
than 70 percent of such gas going up in toxic smoke continuously. Businesses seeking to utilise a bit of
the flared gas are now lining up to be listed as CDM projects and thus eligible to reap profits from carbon
credits.

best seen in the context of public relations. The reality is that these companies are busy raising hurdles on the path of halting the criminal activity. They have demonstrated this in two key ways.

Shell, ExxonMobil and Chevron are said to be deliberately frustrating government efforts to install real-time measurement equipment at 166 gas points to accurately meter the amount of gas being produced in the country. According to the Directorate of Petroleum Resources (DPR) only ten out of the 166 points have had the measuring equipment installed. This posture compounds the lack of transparency in the Nigerian oil; and gas sector, where the true amount of crude oil extracted in the oilfields remains a mystery. More power plants would have come on stream by now, but investors complain that the flaring oil majors have generally refused to cooperate with them and deny them access to the gas that is currently being flared. In fact, three years after thirteen companies prequalified by the Nigerian government to gain access to and harness gas from 180 identified onshore and offshore flare sites; the oil companies have not granted these companies the needed access.

Secondly they are vigorously resisting the moves by the government to bring in a Petroleum Industry Bill (PIB) that would demand more transparency as well as more socio-economic justice in oil and gas sector operations.

By the close of the legislative session that ended in May 2011 there were seven versions of the PIB in circulation to make the law pliable to the desires of the contending forces. In fact, there was only one mention of gas flaring in the version of the PIB that civil society groups believed was likely to be passed into law at that time. Shockingly, the single mention categorised gas flaring as one of the needed indices for a community to expect revenue aimed at community development. The draft PIB said nothing about stopping gas flaring.

The World Bank states that gas flaring decreased in 2009 in Nigeria from 21.3 billion cubic metres to 15.2 billion cubic metres.[3] However, one of the major offenders, the Shell Petroleum Development Company (Shell), admitted that their flares increased 33 per cent in 2010 over their 2009 figure. This clearly shows that whatever may have been the decrease in 2009 was not likely a result of the activities of the oil companies to curb the practice.

Oil companies in Nigeria claim that gas flaring became standard industry practice from the onset of oil exploitation in Nigeria because of a lack of market for the gas. There is a huge market now, but the companies are still lighting up new flares. For example, Shell lit a new one at Opolo-Epie in 2010 although it has been off for a few months now. They also lit another one at their Central Oil and Gas Processing Facility (CPF) at Gbaran-Ubie in the same

3. Statistics on gas flaring in Nigeria are notoriously unreliable. Sometimes the ratio of gas flared is computed without taking into account that the complaint is about gas associated with crude oil extraction and not necessarily from purely gas fields. With more gas from gas fields being utilised for power generation and other uses one could get the impression that gas flaring had reduced whereas that may not be the case. The best overall measure of gas being flared in Nigeria can be seen at the Gas Flare tracker site, http://gasflaretracker.ng (accessed 7 June 2016)

year. Interestingly, an environmental study ordered by Shell gave an interesting verdict passing off gas flaring as having health and environmental benefits. One of the findings of the evaluation was that the incidence of malaria in the area reduced from 29.1 per cent to 26 per cent after their flare lit the sky.

AGIP, the Italian oil giant, however, commenced flaring at their location at Ondewari by mid 2011. Like Shell's flares at Oben and elsewhere, this flare is aligned horizontally at ground level, at Ossiama Creek in Bayelsa State, Nigeria. The company works under heavy guard mounted by the Nigerian Joint Military Taskforce (JTF) and community people passing by in canoes or boats are forced to keep their hands in the air whenever they are close to the location. The poor people are humiliated so that the oil companies and the Nigerian government may continue to reap profits from the unabated assault on the environment and on the people.

Some gas flares have gone off due to the coming on stream of power plants in the Niger Delta. Even where flares have gone out, like at Shell's facilities at Imiringi (Kolo Creek), Etelebou and JK4, these are still like drops in the bucket when taken with the overall prodigious flaring in the region.

Another reason why flaring drags on is the fact that the United Nations Framework Convention on Climate Change (UNFCCC) accepts gas-flare-to-power plants as Clean Development Mechanism (CDM) projects. Projects so far accepted as CDM projects include one by AGIP at Kwale and another by Pan Ocean at the Ovade-Ogharefe oil fields in Delta State.[4]

Analysts believe that the carbon reduction claims of the oil companies are grossly exaggerated and that the power projects are aiming to utilise more of gas from purely gas fields rather than gas associated with crude oil extraction. The reason for this is that associated gas is more expensive to harness than non-associated gas.

The World Bank has backed other projects that ride on false claims such as this. An example is the West African Gas Pipeline project that was hyped as key to halting or at least massively reducing gas flaring in the Niger Delta.It is now believed that only twenty per cent of the gas conveyed by the pipeline is associated gas.[5]

While the gas-to-power plants are essential in an electricity-poor nation like Nigeria, we believe it is immoral, unethical and illegal for the UNFCCC to accord these projects a CDM registration when all they do is partially halting an already illegal activity— with no additionality whatsoever. The additionality requirement for CDM projects demands that such projects preform functions that would otherwise not have been carried out if the project had not been

4. See list of registered CDM projects in Nigeria at http://climatechange.gov.ng/division/mitigation/cdm/registered-cdm-projects-in-nigeria/ (accessed 29 May 2016)

5. Before completion of the pipeline it was projected that 60 percent of the gas it would convey would be associated gas. See for example http://www.energytribune.com/591/west-africa-gas-pipeline#sthash.NslQyyt9.dpbs accessed 29 May 2016

executed. Certainly the World Bank and the various arms of the United Nations, including the UNFCCC must be aware of the illegality of gas flaring.

It is reasonable to assume that even drunks have moments of sobriety. At such times they are able to act without the influence of alcohol. The world has been drunk on crude oil, but considering the sheer harm done to people living in the backwaters of the oil fields and the general harm to the planet, the time is ripe for the world to step back, reflect and halt the harmful practice of gas flaring.

9

How would you fly to the UK?

This piece was written in February 2011 and published in 234NEXT online newspaper, Lagos. It was written in response to fossil fuel addicts who criticise climate justice advocates for flying[1]

The World Social Forum kicked off on Sunday, February 6, [2011] with a march on the streets of Dakar, Senegal. Among the thousands that marched under the careful watch of the Senegalese military and police, were people calling for support for the popular actions in Tunisia and Egypt. There was palpable feeling of invigorated possibilities of globalising peoples' power.

I walked behind a banner with the phrase 'Leave the Fossil Fuels in the Soil' closely followed by another that demanded, 'Do Not Incinerate Africa'. A couple of days later, I posted the photo with the banner on the web. Within minutes, I got a response from a friend who asked, If we leave the oil in the soil, how would you fly to the United Kingdom?'

That question required not just a response, but additional questions. Why must I fly to the UK? Is flying the sole reason for the large-scale environmental assault on poor communities that follow oil extractive activities? Does the ease of my flying to the UK warrant the human blood embedded in every barrel of oil that circulates around the world today from the oil fields of Iraq, Nigeria, and elsewhere? Are we serious about combating climate change if we are not ready to change ourselves, the way we think, the way we produce, and the way we consume?

As I reflected on these questions while participating in climate justice debates at the ongoing World Social Forum, I could not help but ponder on the nexus between crude oil extraction, dictatorship, the scramble for Africa, and the unfolding events in Egypt and the global response.

We have seen the hesitation of major world powers to denounce the clinging on to power by the Pharaoh who has been ruling Egypt over the past three decades. Should we expect support for the popular struggle for peoples' freedom to choose who leads them, or would the world powers merely

---

1. http://news2.onlinenigeria.com/news/general/71740-OIL-POLITICS-How-would-you-fly-the.html (accessed 29 May 2016)

move to ensure that the crude oil movement from and through Egypt remains unimpeded?

Although Egypt is a major player in oil production in the Mediterranean Sea fringe, her strength over the crude oil business globally is due to her control of the Suez Canal, which provides a short link between the Arabian oil fields and Europe.

While Nigeria and Angola top the charts of oil production figures in Africa, Egypt has had steadily rising oil reserves profile, especially with finds in the deep water off the Mediterranean coast. At a point, the country's reserve was said to reach 8.2 billion barrels of crude oil and some 60 trillion cubic feet of gas.

Up to 3,000 oil tankers are said to pass through this canal every year. Besides helping oil vessels make a short ride to Europe and elsewhere through the Suez Canal, Egypt also runs a 320-kilometre-long oil pipeline that goes from the Ain Sukhna terminal by the Red Sea to Sidi Kerir on the Mediterranean coast which 2.5 million barrels of oil pass through daily.

Oil stokes fire

Among the many mineral and other resources that Africa boasts of, and which have stoked the fires of conflict on the continent, oil stands out. Many African countries continue to suffer violent conflicts, human rights abuses, and political instability because of forces struggling to control the oil fields and the associated wealth.

Oil has played a major role in the exploitation and suppression of the peoples of South Sudan and so their eagerness to draw away from the North is understandable. But even after political separation, the two Sudan will nevertheless remain tied together by an umbilical cord of oil pipelines and related infrastructure.

Ghana became an oil exporter in 2010. For the first time, cocoa and gold will face a serious challenge as top income earners. But, just as the government expects huge revenues, the people of the territory where the oil is being extracted are already worried about the expected impacts[2]. Last month, after the first oil export, at least four oil spills have been recorded. Is Ghana echoing the Nigerian situation?

In East Africa, Uganda planned to commence commercial extraction of crude oil in the last quarter of 2011. The oil is drilled in protected areas along the coast of Lake Albert in the famous Rift Valley area. It is a potentially explosive enterprise as the lake is shared by Uganda and the Democratic Republic of Congo; an oil spill here will likely affect both countries.

---

2. Ghana Oil Watch.15 May 2013. Oil Production Threatens Fish Catch. http://ghanaoilwatch.org/index.php/ghana-oil-and-gas-news/3088-oil-production-threatens-fish-catch (accessed 7 June 2016);
   See also a concise report 'oil versus fish, Ghana' at https://ejatlas.org/conflict/jubilee-field-oil-versus-fish-ghana (accessed 7 June 2016)

Moreover, this lake is the source of River Nile and an oil spill here will impact Sudan and Egypt downstream. It has already generated human rights abuses such as restriction of movement in the area and threats of detention by security forces.

Africa is literally awash with crude oil and crude oil addicts are strategising on how to sink their teeth into the waiting veins of land.

How would I fly to the UK if fossil fuels were left in the soil? What will the world do when the oil runs dry? As a Saudi Arabian minister once said, 'the Stone Age did not end for lack of stone and the crude oil age will not end for lack of crude oil.' We must check our fossil fuels mentality for the future of humanity.

10

Can Cancún?

This article was written in November 2010 during the Conference of Parties (COP) to the United Nations Framework Convention on Climate Change at Cancún[1]
While welcoming delegates to the Conference of the Parties (COP) of the United Nations Framework Convention on Climate Change (UNFCCC), President Felipe de Jesus Calderon Hinojosa of Mexico said that climate change has been driven by changes in human behaviour, and that a shift in another direction is needed to reverse the trend.

He intoned that the world must embark on the pursuit of 'green development' and 'green economy' as the path to sustainable development.

He also stated that some of the steps to be taken to attain this ideal include progress on the negotiations on Reduced Emissions from Deforestation and Degradation in Developing Countries (REDD), as well as development of technologies to reduce fuel emission.

These were nice words. These were also very contentious ideas. There are several red flags and concerns about REDD by indigenous groups and forest dependent peoples, as well as mass social movements across the world. The idea of canvassing the extension of financial assistance to the poorest and the most vulnerable countries is also seen by critics as a possible way of dividing them and making them pliable to suggestions and decisions that may actually be contrary to their best interests.

Even before the Cancún conference opened, there were concerns that efforts may already be afoot to rig the outcome, as was the case in Copenhagen in 2009. One concern is about a text for negotiation that is emanating from the chair of one of the working group through an opaque process.

Another concern has arisen from a decision of the Mexican president to invite selected heads of states to the conference. The list is not openly available, but already it is becoming clear that some uninvited presidents intend to be in Cancún.

Last year in Copenhagen, the COP began and ended under a cloud of doubts and perceived undemocratic actions. At that meeting, many delegations

---

1. http://news2.onlinenigeria.com/news/general/59295-OIL-POLITICS-Can-Cancun.rss (accessed 31 May 2016)

from developing and vulnerable nations believed that drafts of what would be the final outcome document were being discussed and circulated within privileged circles, away from the standard practice where such negotiations take place on the open conference floor.

In Copenhagen, there was a steady flow of leaked documents allegedly prepared by the president of the COP. The anxiety in Cancún is being raised by the texts prepared by the chair of the ad hoc working group on Long-term Cooperative Action (LCA). The other major working group under the COP is the one that deals with the Kyoto Protocol and another text is being expected from the chair of that working group, also without a mandate from the working groups, according to analysts.

The year between conferences is spent in technical negotiations and preparations during which delegations review texts prepared by chairpersons of the working groups on the basis of the submissions made by the delegations or members.

The document produced by the chair of the LCA appears to be something quite at variance with what many delegates expected would be the outcome of the negotiations and work done since Copenhagen. The document that delegates is to debate is allegedly based on the 'Copenhagen Accord', which some delegates insist was not an agreement at the end of COP15, but was merely taken note of by that conference.

Questions are being asked why such a document would now be legitimised and made the foundation for serious negotiations expected to produce a fair and ambitious agreement at the end of the conference in Cancún.

After the Copenhagen conference ended without an agreement, the government of Bolivia hosted a first ever World Peoples Conference on Climate Change and the Rights of Mother Earth in Cochabamba in April 2010. The outcome of that conference was the Peoples Agreement that the government of Bolivia then articulated into a formal submission to the UNFCCC and the secretary general of the United Nations.

The essential fault line between those following the path crafted by the Copenhagen Accord and those who do not accept it as the way towards fair agreement that recognises the principle of common and differentiated responsibilities, are quite serious, and the resolution has deep consequences for the future of our planet and the species that inhabit it, including humankind.

The draft text circulated by the chair of the LCA puts forward the ambition that may lead to an aggregate global temperature increase of up to 2 degrees Celsius, as opposed to proposals made by a number of delegations that the target should be between 1 degree and 1.5 degrees temperature rise above pre-industrial levels. A 2 degrees Celsius temperature increase would mean catastrophic alteration to some parts of the world, with Africa being particularly vulnerable.

The text in question has also disregarded the demand by vulnerable nations that to ensure urgent and robust technology transfer for the purpose of mitigation and adaptation, such transfers should not be governed by subsisting intellectual property rights regimes.

Another sore point in the text is that the financial commitment proposed does not step up to the level of ambition needed to tackle the climate crisis, and is even less serious than what was suggested by the so-called Copenhagen Accord.

The immediate past chair of the COP in her final statement indicated that the conference must move in a way that would show that Cancún can deliver a good outcome for tackling climate change. At the end of the first day, the clear question on many minds was, can Cancún?

11

The betrayal of Cancún

This piece was written in December 2010 at COP16 at Cancún and was published by Friends of the Earth International[1]

It was obvious to observers that the climate negotiations at Cancún were wired to support commerce rather than tackling the climate crisis that the world is confronted with. This trend took solid steps a year earlier at the summit in Copenhagen when a handful of nations sidestepped the multilateral tradition of the United Nations and working through 'green rooms' away from the conference floor concocted the so-called Copenhagen Accord instead.

The Copenhagen Accord could not be adopted at the end of the 2009 conference for the basic reason that majority of country delegations did not know how it was crafted and on what basis. Countries like Bolivia and Venezuela stood resolutely against it and that conference only agreed to take note that such a document existed.

The fact that the Accord was not adopted as a conference outcome did not deter its authors, principally the United States, from working behind the scenes, bilaterally, to get several countries to endorse it. Some analysts have said that the endorsement was achieved through arm-twisting tactics and promises of financial and other aids. Those who refused to yield were sanctioned by way of having climate or environment assistance cut.

From the beginning of the Cancún negotiations, signals were sent that its essence was to elevate the Copenhagen Accord to the level of being the conference outcome. The first salvo was fired by the delegation of Papua New Guinea who declared that a few nations with divergent votes from the majority must not prevent the conference from reaching a decision. They suggested that if a consensus became impossible a decision should be made by a vote. This position, as noted in an earlier article on Cancún, was immediately objected to by the delegations of Bolivia, India, Saudi Arabia and others.

At the end of the Cancún summit, with the Copenhagen Accord now dressed in new garbs, there was no consensus for its adoption. Not to be deterred, the Mexican presidency of the conference banged the gavel repeatedly

---

1. http://members.foei.org/en/blog/2011/01/03/the-betrayal-at-cancun on 3rd January 2011 (accessed 31 May 2015)

on her table and rammed the document through, after redefining consensus as not necessarily meaning unanimity.

Empty promises

Nations yelped and cheered. Cancún had delivered; they enthused and backslapped each other. But what did Cancún deliver and how will the planet fare under the scenario set by what has been termed Copenhagen Accord 2?

The conference outcome avoided legally binding emissions reduction targets for the main polluting nations— the rich industrialised countries— and rather urges a voluntary pledge-based system with no monitoring mechanisms. From recent WikiLeaks revelations regarding discussions in France, it is clear that the rich countries are determined not to make binding commitments to act for the safety of the planet.

Looking for something to celebrate, some countries latched on the promise to create a Climate Fund within the United Nations climate change framework but having the World Bank as a trustee. The promised climate fund did not specify how the funds would be sourced.

The agreement did not review subsisting intellectual property regime that does not freely allow the exchange of green technology. It took big steps in paving the way for new market based mechanisms that would allow for speculation and avoidance of actions to reduce emissions at source and generally position the planet at great risks of catastrophic climate change.

Teresa Andersen of the Gaia Foundation, who wrote about the manner the Cancún conference[2] ended, captures the disbelief of critical observers:

> We sat in disbelief as the crowds leapt to their feet, cheering, applauding, whooping and whistling the Mexican chair of the Cancún climate negotiations. Mexico's foreign secretary, Patricia Espinosa, graciously bowed her head, her hands crossed over her heart in an authoritarian simulation of modesty, as we shook our heads, open-mouthed, at the eerie frenzy taking place around us. In the last hours of the Cancún climate negotiations, the world's deluded leaders were cheering as they tossed the planet onto the bonfire.

According to Teresa,

> The Cancún Agreement, we are told, has 'saved multilateralism'. What it has not done though, is offer any meaningful solution to climate change. As it stands, the Cancún Agreement could mean

2. Teresa Turna.21 December 2010. Mass-Hypnotism in Cancún Climate Delusion – where 'saving multilateralism' means burning the planet. http://www.gaiafoundation.org/blog/mass-hypnotism-cancun-climate-delusionaccessed 31 May 2016

global temperature rises of up to 5 degrees centigrade, and a possible 6.5 degrees in Africa.

An initial analysis of the Cancún outcome by Friends of the Earth International (FOEI) saw the prospects of opening new market mechanisms as potentially creating practices that are more harmful to the climate than current ones.

According to FoEI,

> ...the establishment of one or more market-based mechanisms over the course of the next year is to be considered, with a view to taking a decision to adopt these new mechanisms at COP17 in South Africa. The new mechanisms could include a number of different types of instruments, some of which would be more destructive than others.

All was not lost in Cancún. Social movements pushed the path of climate justice in various venues in Cancún. The government of Bolivia, which had facilitated a Peoples Conference on climate change and the Rights of Mother Earth in April 2010, stood with the people, pushing the right analysis and solutions, right to the end of the conference.

Social and climate justice movements clearly stated that the causes of climate change are systemic and that the only way to tackle the climate crisis is through a change of the capitalist and patriarchal system that caused it.

With the clear indication that rich nations are not keen to tackle climate change, but would rather make bogus promises that poor vulnerable nations unfortunately lap up, it is doubtful if the 2011 conference to be hosted in Durban, South Africa, will produce anything different from Copenhagen and Cancún.

The South African government has dubbed COP17 the Peoples COP. It will be seen whether the voices of the people will prevail or if corporations and their surrogate politicians will hold sway in their market-based chariots.

12

A red card for California REDD

This article is from a blog written about California REDD on 5 July 2013[1]
The state of California, USA, has become the battleground for REDD-type projects. REDD is the acronym for Reducing Emissions from Deforestation and Forest Degradation. While there is no argument against deforestation and forest degradation, many critics and forest-dependent communities literally see red in the practical implications of REDD as a tool to combat global warming.

California is on the verge of allowing carbon credits obtained from forests and tree plantations anywhere in the world to be used as offsets for polluting activities at home. Targeted forests include those in Acre, Brazil as well as in Michigan in the USA.

In bringing up the UN-REDD Framework, the United Nations admitted that REDD could result in the 'lock-up of forests', 'loss of land' and 'new risks for the poor.'[2] The No REDD in Africa Network (NRAN) in a recent statement rejected the inclusion of REDD projects in the State of California's Global Warming Solutions Act, AB32. NRAN stated that just as the 'UN predicted, in Africa, REDD and forest carbon projects are already resulting in 'loss of land' in the form of massive evictions, as well as 'new risks for the poor' in the form of servitude, slavery, persecutions and killings.'[3]

Indigenous groups in Brazil and Mexico as well as NRAN and Oilwatch International have sent petitions to the Governor of California, the Chairman of California Air Resources Board and other officials of the California

1. http://nnimmo.blogspot.com/2013/05/red-card-for-california-redd.html and in *The Africa Report* (08 July 2013) at http://www.theafricareport.com/Columns/big-red-card-for-california-redd.html (accessed 31 May 2016)

2. Lang, Chris (2013), 'There is no safe REDD': Global Alliance of Indigenous Peoples and Local Communities on Climate Change against REDD and for Life. http://www.redd-monitor.org/2013/05/10/there-is-no-safe-redd-global-alliance-of-indigenous-peoples-and-local-communities-on-climate-change-against-redd-and-for-life/ (accessed on 31 May 2016)

3. REDD Monitor. 4 May 2013. No REDD in Africa Network opposes inclusion of REDD offsets on California's cap-and-trade scheme. http://www.redd-monitor.org/2013/05/04/no-redd-in-africa-network-opposes-inclusion-of-redd-offsets-in-californias-cap-and-trade-scheme/ (accessed 31 May 2016)

Environmental Protection Agency, demanding the exclusion of REDD from California's climate solutions.

The spread and diversity of the groups standing against California REDD stem from the fact that this may unlock an avalanche of REDD-type projects around the world. These projects would operate both outside and within the UN-REDD system. The implication according to Oilwatch International is that polluting companies such as Shell could continue polluting while imagining that their carbon emissions are offset by the carbon stored in trees in Brazil, Mexico, USA or Nigeria.

Critics see REDD as a dangerous false solution to global warming primarily because it locks in pollution. It also locks out communities from their forests, impacts on their culture and blocks off their sources of livelihood. REDD also does not halt deforestation but at best displaces this objectionable act to another location or merely delays it. Carbon offset projects exploit forests as mere carbon sinks.

Shell oil company recently purchased 500,000 carbon offsets credits from a forestry project on over 200,000 acres in Michigan 'that not only will grant Shell's refinery in Martinez, California permission to pollute, but will push the planet further down the road to catastrophic global warming', according to Oilwatch International.

In Mozambique, La Via Campesina found in a study on the N'hambita REDD project in Mozambique that thousands of farmers were paid meagre amounts for seven years for tending trees. Because the contract is for ninety-nine years, if the farmer dies his or her children and their children must tend the trees without any further pay or compensation.[4] This has been interpreted as a clear case of carbon slavery. Regrettably, the N'hambita project was celebrated by the UN Sustainable Development website.[5]

Violent evictions in Uganda saw over 22,000 farmers with land deeds shoved off their land for a REDD-type project in 2011. In one of the incidents a sick eight-year-old Friday Mukamperezida was killed when his home was razed.[6]

4. Mozambique : Carbon Trading and REDD : farmers 'grow' carbon for the benefit of polluters: http://www.viacampesina.org/en/index.php/actions-and-events-mainmenu-26/-climate-change-and-agrofuels-mainmenu-75/1265-mozambique-carbon-trading-and-redd-farmers-grow-carbon-for-the-benefit-of-pollutersaccessed 31 May 2016REDD Monitor Envirotrade's carbon trading project in Mozambique: 'The N'hambita experiment has failed.' (accessed 31 May 2016)

5. The N'hambita Community Carbon Project. https://sustainabledevelopment.un.org/index.php?page=view&type=99&nr=32&menu=1449 (accessed 31 May 2016)

6. The Guardian, (2011) *Ugandan farmer: 'My land gave me everything. Now I'm one of the poorest'* http://www.guardian.co.uk/environment/2011/sep/22/uganda-farmer-land-gave-me-everythingaccessed 31 May 2016New York Times, (2011) *In Uganda, Losing Land to Planted Trees* – Slide Show http://www.nytimes.com/slideshow/2011/09/22/world/africa/22uganda-3.htmlaccessed 31 May 2016New York Times, *In Scramble for Land, Group Says, Company Pushed Ugandans Out* http://www.nytimes.com/2011/09/22/world/africa/in-scramble-for-land-oxfam-says-ugandans-were-pushed-out.html?_r=1 (accessed 31 May 2016); REDD Monitor, Ugandan farmers kicked off their

In Nigeria REDD is already raising the spectre of persecution and criminalisation of activists, including in Cross River State, Nigeria where the State of California intends to have REDD projects. The Executive Director of the Rainforest Resource and Development Centre (RRDC) in Cross River State, Nigeria, Mr Odey Oyama, had to flee his home for several weeks in January and February 2013 due to harassment and intimidation from state security agents. Odey is one of the vocal opponents of REDD activities aimed at extracting more forest estates from indigenous communities and similar land grab operations.

> One of the activities placing me in confrontation with the Cross River State Government of Nigeria is my stand against the REDD programme. My reason for rejecting the REDD programme is because it is geared towards taking over the last vestiges of community forest that exist in Cross River State of Nigeria, declared Mr Oyama.

Land grabbing for plantation agriculture in Ogoni land, already decimated by pollution from the oil industry, has turned violent. The Government of Rivers State of Nigeria forcefully seized and gave away over 2,000 hectares of community farmlands in Tai and Khana Local Government Areas to a Mexican company, Union De Iniciativa S.A. de C.V., for the cultivation of bananas possibly for export. A new report issued by the group Social Action[7] indicates that at least three youths have been killed in relation to this land grab. The project was approved and has commenced without an environmental impact assessment as required by law.

In other parts of Africa, REDD is exacerbating threats to the cultural survival of Indigenous Peoples. According to 'The DRC Case Study: The impacts of carbon sinks of Ibi–Batéké Project on the Indigenous Pygmies of the Democratic Republic of Congo' published by the International Alliance of Indigenous and Tribal Peoples of the Tropical Forests, Batwa Pygmies suffer 'servitude'[8] on the World Bank's Ibi–Batéké Carbon Sink Plantation.[9] This REDD-type forest carbon plantation for fuel wood and charcoal is the

---

land for New Forests Company's carbon project http://www.redd-monitor.org/2011/09/23/ugandan-farmers-kicked-off-their-land-for-new-forests-companys-carbon-project/#more-9681 (accessed 31 May 2016)

7. See Killing for Banana- Government Land Grab, Violence and the Forgotten Rights of Ogoni Farmers. http://saction.org/books/Killing_for_Banana_2013.pdfaccessed 31 May 2016

8. International Alliance of Indigenous and Tribal Peoples of the Tropical Forests, 'Indigenous Peoples and Climate Change: Vulnerabilities, Adaptation, and Responses to Mechanisms of the Kyoto Protocol' (2007) Makelo, S., 'The DRC Case Study: the impacts of carbon sinks of Ibi–Batéké Project on the indigenous Pygmies of the Democratic Republic of Congo' p.45-74 especially 62-64The human rights violations against Pygmies are grave throughout the country. See 'Pygmies beg UN for aid to save them from Congo cannibals.' See http://www.genocidewatch.org/images/
D.R._Congo_23_May_03_Pygmies_beg_UN_for_aid_to_save_them_from_Congo_cannibals.pdf (accessed 31 May 2016)

DRC's first Clean Development Project and claims to contribute to sustainable development and climate change mitigation.[10] Pygmy leaders object to these projects and have denounced the World Bank for funding deforestation of their ancestral forests that not only releases emissions but also violates their rights, destroys their livelihood and causes social conflict.

In supporting the protest against California REDD from Brazilian organisations, several groups and movements from around the world stated in an open letter 'We believe that their demand for a *meaningful* participation in any consultation process in Acre related to legislation or programmes linked to REDD activities that already or potentially affect their way of life is legitimate. Forest-dependent peoples have the right to give or withhold their consent to activities that deeply interfere with their way of living. Decisions regarding REDD legislation or programmes already do and will in future affect forest peoples' way of life. Given that such meaningful participation was absent from REDD processes in Acre or during the elaboration of recommendations to the government of California in this matter, we urge you not to include REDD offset credits into the California carbon trading scheme.'[11]

That is a real red card for California REDD.

9. World Bank 'DRC IbiBateke Carbon Sink Plantation' http://wbcarbonfinance.org/ Router.cfm?Page=Projport&ProjID=43647 (accessed 31 May 2016)

10. Reuters: World Bank to buy carbon credit from Congo Project http://www.reuters.com/article/ environmentNews/idUSTRE57409I20090805 (accessed 31 May 2016)

11. Declaration of solidarity with the letter from Acre, Brazil opposing REDD offset credits in California's cap-and-trade scheme. http://www.redd-monitor.org/2013/05/01/declaration-of-solidarity-with-the-letter-from-acre-brazil-opposing-redd-offset-credits-in-californias-cap-and-trade-scheme/ (accessed 31 May 2016)

13

Ambition, selfishness and climate action

This article is from a keynote address presented at the Intensive Practical Course on Climate Change, Environmental Law, Regulation and Management hosted by The Nigerian Institute of Advanced Legal Studies (NIALS) at the University of Lagos, 20 May 2013[1]

Humanity's fossil-fuel addiction will be the climate hangman unless we quickly wean ourselves off them and take a new energy trajectory. We make this assertion because evidence continues to mount and all witnesses – conservative and radical- point unwavering fingers at the oil and gas wells, coal holes and the tar sand pits of this world.

Global warming occurs due to increasing concentrations of greenhouse gases in the atmosphere. Heat comes from the sun in short waves, but when bounced off the earth they go up in long waves. Whereas the short waves pass through the atmosphere without resistance, the greenhouse gases trap some of the long waves trying to exit the atmosphere. Scientists estimate that without the greenhouse effect the earth would be as cold as minus 18 degrees Celsius.That does sound like we should celebrate the greenhouse gases in the air. Right? Well, the trouble kicks in when the concentration of the greenhouse gases gets higher than it ought to be.

The principal greenhouse gas is carbon dioxide (CO_2). Its level has increased by a third since the industrial revolution while that of methane has doubled. Over the past 150 years, a period during which fossil fuels have become the main source for energy needed for electricity generation and or movement of goods and people, temperature have risen by 0.8degrees C and is set to gallop with increasing concentrations of greenhouse gases in the atmosphere.

As the reports of three conservative agencies have shown, unless the world embarks on a swift energy transition the course is set for calamitous global warming. Rather than seek ways to move from dependence on fossil fuels, rich nations are literally fighting to secure fossil fuels reserves to ensure they can keep guzzling and not change their high consumption lifestyles. So-called

1. http://nnimmo.blogspot.com/2013/05/ambition-selfishness-and-climate-action.html (accessed 31 May 2016)

emerging nations, like those in the BRICS bloc are equally ramping up their consumption levels as they assert the right to pollute so as to grow or as they grow. Thus growth may now be seen as a measure of pollution. Researchers estimate that at the current rate of consumption of petroleum resources, China alone can exhaust the known stock in just one decade. Indeed, it is estimated that by 2030 the USA and China will together generate 45 per cent of global carbon emissions.

Fossil fuels have supplied energy more efficiently than most other sources. The fuels have been cheaper than others because the environmental costs are externalised to poor communities and peoples whose governments are satisfied with rents from the sector and could not care less about the impunity in the fields and mines. As fossil fuels resources dwindle, we witness more desperate exploration. We see extraction in protected areas including the Arctic region. We see more aggressive moves into deeper waters and open and blatant warfare conducted in the guise of securing democracies.We can expect a spread of extreme extraction such as is already seen in fracking and tar sands exploitation.

Fracking is short for hydraulic fracturing – a process of blasting a solution of water, sand and a cocktail of chemicals into a shale bed between two to three kilometres into the belly of the earth. The mixture fractures the rock and releases the gas through bruises created by the sand. Because of its sheer depth some of the wells must puncture through aquifers causing pollution either from the chemicals used or as fallout of the violent fracturing process itself.

The pollution and the global warming threats notwithstanding, the race to squeeze the last drops of fossils from the earth is on. An official US Department of Energy Report is quoted to have said 'The world has never faced a problem like this. Without massive mitigation more than a decade before the fact, the problem will be pervasive and will not be temporary. Previous energy transitions were gradual and evolutionary. Oil peaking will be abrupt and revolutionary.'

Leaving the fossils in the ground is the unmistaken path that we ignore to our peril. The United Nations Environment Programme (UNEP) issued a report late 2012 that asked some pointed questions. The questions were premised on the fact that the aggregate voluntary emissions reductions by rich, industrialised and polluting nations would not ensure the level of reduction needed to avoid catastrophic global warming. It showed that a gap existed between the 'level of ambition that is needed and what is expected as a result of the pledges.'

Previous assessments showed that for global temperatures to stay within a 2 degrees Celsius increase annual emissions ought to average 44 gigatonnes (GT) or less by 2020. UNEP scientists, however, believe that current levels are 14 percent above what should be the level in 2020 and that if urgent actions are not taken an emissions gap of 8 Gt of CO_2 equivalent could happen.

Other analysts like Pablo Salon believe that 'With the Doha, Durban and Cancún outcomes they will hit the level of 57 GT of CO_2e by 2020. So the 'gap' is 13 GT of CO2e.' Solon warns, 'If this 'gap' is not closed by 2020 the global average temperature of the planet will increase by more than 4 to 8 deg C. The last time the Earth had a global warming like this was millions of years ago.'[2]

Just before the 18th Conference of Parties of the UNFCCC at Doha in 2012, similar research findings emerged from three unexpected quarters: the World Bank, the International Energy Agency and the business outfit PricewaterhouseCoopers (PwC). The PwC report sees the safe limit of 2 deg C temperature increase suggested by the IPCC as unrealistic because even if current 'decarbonisation' levels are doubled the world would still be heading for a temperature increase of 6 deg C by the end of the century.

Both the World Bank and the IEA reports suggest that for a 50-50 chance of staying below two degrees, the world must leave 66 percent of the known reserves of coal, oil and gas underground. And for an 80 percent chance, we have to leave 80 percent of those reserves untouched. Despite of all these warnings, political leaders dither and pollutions roar ahead as if there will be no tomorrow. Perhaps they know that there may not be any.

Carbon Tracker, a consultancy outfit, reached a similar conclusion earlier. According to the group, for warming to be kept at 2 degrees C, from 2010-2040 only 565 billion tonnes of CO2 can be permitted to be emitted into the atmosphere. If this is done, there would be a 20 per cent chance of success. However, the known fossil fuels reserves have about 2795 billion tonnes of CO2 of which two thirds is coal, 22 per cent is crude oil and 13 per cent is gas. From these figures they estimate that 80 per cent of the known reserves must be left below the ground if we must hope for a slim chance of keeping temperature increase at 2 deg C.

On 9 May 2013 a record was made when CO2 concentration in the atmosphere reached 400 parts per million (ppm) as measured at Hawaii's Mauna Loa observatory. It has been noted that the last time this level of CO2 was attained was 3-5 million years ago. At that time scientists believe that temperatures were 3-4 degrees warmer than they are today and that sea levels were 5-40 metres higher than we have today. In addition, there was no ice in the Arctic region and there probably were no humans on the planet at that time. The concentration of carbon in the atmosphere before man's romance with fossil fuels stood at 280ppm. In less than two centuries we have dug ourselves into deep fossil holes and marched to the climate precipice.

Look at that sea level in the long gone age. Five to 40 metres! Should the earth experience a 1 metre sea-level rise in the future what would become of the Eko Atlantic currently being built into the sea? Generally, because of the low lying nature of Nigeria's coastal region a sea level rise of a mere 1 metre would

2. Pablo Solon. 23 May 2013. Everyone Must Accept Binding Climate Commitments. http://www.climatechangenews.com/2013/05/02/pablo-solon-everyone-must-accept-binding-climate-commitments/ (accessed 31 May 2016)

mean the inundation of land quite a distance into the hinterland. There is an estimate that this could go as far inland as 90 kilometres.

Temperature rises pose universal problems to the whole world, but more so for Africa. This is so because Africa has 50 percent higher temperatures than the global average. If temperature increases by say 4 deg C, Africa would be 6 deg C warmer. The consequences would be dire. We can expect mass crop failures, concomitant starvation and mass migration for those who can.

At the Copenhagen summit in 2009, the lead negotiator for G77, Lumumba Di-Aping[3], denounced the 2 deg Celsius warming target as 'certain death for Africa' and as a type of 'climate fascism' forced on Africa. Di-Aping then said Africa was asked to sign an agreement that would permit warming in exchange for US$10 billion, and that Africa was also being asked to celebrate that deal.

Michael Mann, speaking on Democracy Now! warned, 'We have to go several million years back in time to find a point in Earth's history where CO2 was as high as it is now. … If we continue to burn fossil fuels at accelerating rates, if we continue with business as usual, we will cross the 450 parts per million limit in a matter of maybe a couple of decades. With that amount of CO2 in the atmosphere, we commit to what could truly be described as dangerous and irreversible changes in our climate.'[4]

Warning that we were trudging on the roadmap to idiocy, George Monbiot looked at the atmospheric pollution record and suggested a possible way out of the fix: 'The only way forward now is back: to retrace our steps and seek to return atmospheric concentrations to around 350ppm, as the 350.org campaign demands. That requires, above all, that we leave the majority of the fossil fuels which have already been identified in the ground. There is not a government or an energy company which has yet agreed to do so.'[5]

In the face of the clear warnings, oil companies and others benefiting from the world's fossil fuels addiction continue to press on unperturbed. Again we turn to Monbiot's blog: 'Recently, Shell announced that it will go ahead with its plans to drill deeper than any offshore oil operation has gone before: almost 3km below the Gulf of Mexico. At the same time, Oxford University opened a new laboratory in its department of earth sciences. The lab is funded by Shell. Oxford says that the partnership 'is designed to support more effective

3. Cory Morningstar. 10 December 2012. The Most Important COP Briefing that No One Ever Heard- Truth, Lies, Racism & Omnicide. http://climatesoscanada.org/blog/2012/12/18/the-most-important-cop-briefing-that-no-one-ever-heard-truth-lies-racism-omnicide/ (accessed 31 May 2016)

4. Democracy Now!.13 May 2013. Climate Tipping Point? Concentration of Carbon Dioxide Tops 400 ppm for First Time in Human History. http://www.democracynow.org/2013/5/13/climate_tipping_point_concentration_of_carbon (accessed 31 May 2016)

5. George Monbiot. 10 May 2013. Climate milestone is a moment of symbolic significance on road of idiocy. http://www.theguardian.com/environment/georgemonbiot/2013/may/10/carbon-dioxide-milestone-climate-change (accessed 31 May 2016)

development of natural resources to meet fast-growing global demand for energy.' Which translates as finding and extracting even more fossil fuel.

Have the Conference of Parties helped the world to tackle climate change? The United Nations Framework Convention on Climate Change (UNFCCC) is the space where nations negotiate and should agree to act together in the common interest and for the survival of the planet. The conferences of parties (COP) to the convention have over the years turned into sessions where the powerful browbeat the weak and efforts are made to avoid responsibility and to act in narrow national or regional interest. The rapid slide down this slope took root at COP15 in Copenhagen, got deepened at COP16 in Cancún where the concept of consensus got redefined as agreement by the majority. COP17 in Durban took the medal as a conference whose critical achievement was the blatant postponement of action while the earth burns. Nations like the USA, Canada, Japan and Australia openly throw spanners in the works. Some go as far as foreclosing any participation in any legal and accountability formats proceeding from the Kyoto Protocol.

Doha was a sigh as leaders kicked the noisy decision-making can further down the road. There was little excitement about COP18 at Doha even before it took place and no celebratory vuvuzelas were heard after the event either. In the negotiations following Doha the talks in Bonn and Geneva continue to show the strains between developed, emerging economies and differently developed nations – especially with regard to emissions reductions commitments and mitigation actions.

At the negotiations held early May 2013 at Geneva the developed countries pushed for a legally binding 'spectrum of commitments' from both developed and developing countries. However, their stance was based on targets nationally determined according to national capabilities and circumstances. They suggested that these would be reviewed periodically with the aim of keeping global temperature rise in line with the 2 degree Celsius goal.

Nations dance to different beats as they negotiate. Reporting from the meetings, the Third World Network informed that the developed countries also wanted 'a different form of differentiation according to the emission profiles of countries rather than that which exists in the Convention (which is a differentiation between developed and developing countries). Developed countries also wanted common accounting rules for mitigation and transparency for both developed and developing countries. The United States did not want developing countries to condition their contribution to emission reductions on the availability of finance and technology transfer.'

The position of many of the developing countries was that the differentiation applied must remain the same as in the Convention keeping the lines of developed and developing countries or Annex 1/Non-annex 1 and not diminishing the importance of the historical responsibility of developed countries. They also insisted that the developed countries should take the lead in emissions reductions and for finance, technology transfer and capacity building to be provided to developing countries.

This spat can indeed be seen as the cause of the lethargy underlying the politics in the negotiations and keeping leaders from considering the need for real actions to tackle global warming. The developed nations see any real emissions reductions as potentially slowing their development curve, challenging their industries and ultimately placing heavy financial burdens on their systems and peoples. The developing nations on the other hand insist that developed countries must bear their historical responsibility for taking up as much as 80 per cent of the atmospheric space for carbon. The debates about emissions reductions can in a sense be seen as a struggle about who would colonise the remaining atmospheric space.

Climate justice advocates generally insist that those who created the climate problem must be the ones to mitigate it. However, there is a rising call also from these quarters that some level of binding commitment by developing nations may be in order. Even here, the argument is that the commitments must be based on common but differentiated responsibilities.

Clearly, a bottom-up or voluntary emissions reduction would not work, as nations are unwilling to radically cut their emissions. Developed nations generally choose the market track and rely on offsets to do the mathematics of emissions reduction. Developing nations, including highly polluting and emerging nations like China and others in the BRICS formation prefer to take cover under the umbrella of the developing nations and claim the right to develop as equal to the right to pollute. The point against a bottom up, voluntary path is that there is no mechanism for closing the emissions reduction gap should the pledges not add up to the ambition needed by 2015 to enhance mitigation actions by 2020, etc.

The inability to meet climate finance and adaptation needs in the face of rising military budgets give us deep instructions. When you are standing at the precipice, you do not make progress by stepping forward. The effective action is stepping backwards. At such a time, it would make sense to be content with *Keke* NAPEP[6] going in the right direction than insisting on the luxuries of a stretch limousine heading in the wrong direction, metaphorically speaking. Pressing ahead as we see in the climate talks as well as in some localised actions simply show that humans have refused to accept the evidence around us.

With the crisis on hand, the need to provide adequate finance for climate mitigation and adaptation is more serious than ever before. Yet the UNFCCC processes have merely thrown up a Green Climate Fund with a literally empty kitty. The US$10 billion per year over a three-year period carrot dangled in Copenhagen and the promise to ramp that up to US$100 billion a year from 2020 has not materialised. To refresh our memory, we quote the 'accord' below:

The collective commitment by developed countries is to provide new and

---

6. A trycycle used as taxis in parts of some Nigerian cities

additional resources, including forestry and investments through international institutions, approaching US$30 billion for the period 2010–2012 with balanced allocation between adaptation and mitigation. Funding for adaptation will be prioritised for the most vulnerable developing countries, such as the least developed countries, small island developing states and Africa. In the context of meaningful mitigation actions and transparency on implementation, developed countries commit to a goal of mobilising jointly US$100 billion dollars a year by 2020 to address the needs of developing countries. This funding will come from a wide variety of sources, public and private, bilateral and multilateral, including alternative sources of finance.

Most wars have been fought to secure resources and to expand spheres of influence. This has become even clearer today as resource wars are fought under a variety of false pretences. As the resources get depleted the intensity of the conflicts will increase. And so will the military budgets.

Military expenditure by the industrialised nations went up by 50 percent since 2001 and rose to over US$1.7 trillion in 2011. A mere fraction of that amount would save lives and help combat the ravages of global warming. Somebody figured that if a dollar represented one second it would require 32,000 years to reach 1 trillion. We are talking of huge sums here. If just 25 percent of the war budgets were to be set aside for climate mitigation/adaptation measures there would be US$434.5 billion in the kitty and the world would be the better for it. Just consider that one stealth bomber costs a whopping US$1 billion.

Another source of funds would be for the rich nations to pay for the ecological debt owed the nations and regions that have borne centuries of prodigious exploitation and environmental damage. While debates go on in scholarly circles about how such a debt could be computed and to whom it would be paid, rich nations have simply refused to consider the notion. They insist on staying on the path of limitless and continuous growth. But it is noteworthy that 'an economy based on growth and resource depletion cannot function globally, since it logically implies that power is accumulated in one part of the world and applied in another. It is in essence particularist, not universal: everyone cannot exploit everyone else at the same time.'[7]

Climate change has become big business and false solutions are celebrated just as the naked emperor was hailed as being well dressed. Whereas it has been clear for a long time now that global warming is mostly man-made and is due to the huge amount of greenhouse gases pumped into the atmosphere by polluting activities involving the use of fossil fuels, preferred actions taken by nations and industries have been patently false actions. These actions are mostly predicated on the specious notion of carbon offsetting. The notion itself is built on the creed that financial markets hold the key to solving humankind's problems.

7. Harald Welzer.2012. *Climate Wars: What People Will be Killed For in the 21st Century*. Malden, Polity Press.

Carbon offsets allow polluters to keep polluting provided they pay for it in cash (carbon tax) or imagine that some trees somewhere else in the world are absorbing an equivalent carbon as they are emitting in their activities. Thus while damaging the climate, polluters perform acts of indulgence through offsets.

For example, the so-called Clean Development Mechanism (CDM) covers some of such offset schemes where projects that help reduce carbon emissions earn some carbon credits. Some really obnoxious projects get listed under the CDM. Gas to power projects utilising gas that was otherwise flared make sense, except you consider the fact that gas flaring has been illegal in Nigeria since the gas re-injection law came into effect in 1984. There has also been a High Court judgment in the case of Jonah Gbemre versus Shell Development Petroleum Company over the gas flare at Iwerekhan, Delta State. The High Court sitting in Benin City ruled that gas flaring is an illegal activity, is unconstitutional and is an affront on the people's human rights. That judgment was delivered in November 2005 but the flares continue to roar.The point here is that even if the gas to power plants succeeded in stopping gas flaring, they would simply have helped to stop an illegal activity and should not merit consideration as CDM projects. Qualifying projects are expected to be ones that bring in additionality, or that do some mitigating actions that would not have otherwise been done.Writing on this elsewhere we made the point that 'Any compensation for such an activity flies in the face of reason. Gas flares are the most cynical manifestations of corporate insolence in the face of climate change and environmental health. The flares release greenhouse gases such as carbon dioxide, methane and nitrous and sulphur oxides. Apart from these, the flares release other harmful substances that greatly affect human health.'[8]

Just when we thought we had overcome slavery we are getting dragged away into not just carbon colonialism but carbon slavery. Carbon was placed on the market shelf through the acceptance of the CDM at the COP held in Kyoto in 1997. That opened the floodgates for carbon speculators, introduced inaction and benefited carbon cowboys while disasters hit the world's vulnerable communities and sometimes the rich!

Market mechanisms threw Reducing Emissions from Deforestation and Forest Degradation (REDD) into the tray at the Bali climate meeting of 2009. REDD and its variants allow polluters to keep on at their business of polluting while 'showing' that trees in a forest or plantation that they have secured somewhere else absorb the carbon they emit. Thus REDD projects permit pollution and cannot be said to reduce emissions. It is clear that the name itself is a sad joke. In addition, REDD does not stop deforestation, but at best defers or displaces it. A REDD scheme is a business scheme, pure and simple.

A declaration from the Climate Space at the World Social Forum held

8. Nnimmo Bassey.2012. *To Cook a Continent – Destructive Extraction and the Climate Crisis in Africa.* Oxford- Pambazuka Press.

in Tunis in March 2013 insisted 'We cannot put the future of nature and humanity in the hands of financial speculative mechanisms like carbon trading and REDD. REDD (Reducing Emissions from Deforestation and forest Degradation), like Clean Development Mechanisms, is not a solution to climate change and is a new form of colonialism. In defence of Indigenous Peoples, local communities and the environment, we reject REDD and the grabbing of the forests, farmlands, soils, mangroves, marine algae and oceans of the world, which act as sponges for greenhouse gas pollution. REDD and its potential expansion constitutes a worldwide counter-agrarian reform which perverts and twists the task of growing food into a process of 'farming carbon' called 'Climate Smart Agriculture.'

The Climate Space also opposed 'proposals that want to expand the commodification, financialisation and privatisation of the functions of nature through the so-called 'green economy' which places a price on nature and creates new derivative markets that will only increase inequality and expedite the destruction of nature.'[9]

Groups like the No REDD in Africa Network (NRAN) see REDD as a dangerous false solution to global warming primarily because it locks in pollution, just as the UN-REDD framework feared would be the case when the scheme was introduced. REDD locks out communities from their forests, impacts on their culture and strangulates their sources of livelihood. REDD schemes see forests and plantations as little more than carbon sinks.

Some REDD-like projects operate outside the purview of the UN-REDD coverage. One of such schemes is what has come to be called California REDD. Moves to include REDD projects in the state of California's Global Warming Solutions Act, AB32 has drawn a lot of criticism from around the world because many believe that this would give impetus to similar schemes to mushroom around the world, granting polluters more space to keep on with their harmful activities thereby placing the world in deeper problems. It was in this wise that Oilwatch International denounced Shell oil company's purchase of 500,000 carbon offsets credits from a forestry project on over 200,000 acres in Michigan, USA because it would grant Shell the permission to pollute at its refinery in Martinez, California.

In its own rejection of California REDD, NRAN recalls a situation in Mozambique, where a La Via Campesina study found that thousands of farmers in the N'hambita REDD project were paid meagre amounts for seven years for tending trees. 'Because the contract is for ninety-nine years, if the farmer dies his or her children and their children must tend the trees without any further pay or compensation. This has been interpreted as a clear case of carbon slavery.'[10]

9. Climate Space. Declaration of Climate Space, WSF. https://www.alternatives.ca/en/content/story/declaration-climate-space-wsf (accessed 31 May 2016)

10. Africans Unite against New Form of Colonialism: No REDD Network is Born. http://wrm.org.uy/oldsite/subjects/REDD/Africa_No_REDD_Network.htmlaccessed 31 May 2016

Another false solution has been the presentation of agrofuels as a replacement of fossil fuels. It is a false solution because it keeps the fossil fuels paradigm and is equally polluting. Moreover it has triggered massive land grabs and even at its peak cannot replace fossil fuels because the amount of land needed to cultivate crops and the feedstock needed for production of agrofuels is simply not available on planet earth.

Geo-engineering and agricultural genetic engineering are other false solutions that lull humans to think that they can keep current polluting lifestyles and find techno-fixes for their addiction.

What must be done? Reflections on the challenge of climate change can leave us utterly exasperated considering the corporate capture of governments and the refusal of states to take actions that would benefit the people and the planet and not just the corporations. Although time is ticking fast, the peoples of the world must continually press for climate justice, understanding that no nation, rich or poor, is immune to the challenge of global warming. This has been amply illustrated by the tragic weather events that have fairly democratically impacted nations around the world. These are undeniable:

- Sea levels are rising
- Arctic ice is melting – may lead to changes in ocean circulation
- Sea-surface temperatures are rising
- Acidification of sea water due to increase of dissolved carbon dioxide
- Heavier rainfalls, hurricanes and floods are common
- Droughts and desertification getting more intense
- Crop failures
- All these and more impact negatively on human lives and that of other species on planet earth. Urgent actions are needed across all nations. Among these we list:
- A just global climate treaty that recognises historical responsibility, climate debt as well as legally binding emissions reduction
- Elimination of market mechanisms (including CDM, REDD, REDD) and all other false solutions from the climate regime
- Rapid transition from dependence on fossil fuels— including in transportation, power generation and agriculture
- Recycling of wastes
- Make national laws that build mechanisms for climate mitigation and adaptation actions including coastal protection, combatting desertification
- Stop gas flaring in the Niger Delta and at Badagary immediately
- Stop fracking and other extreme extraction including drilling in the Artic region
- Creation of communities climate defence committees that would set rules for physical developments as well as monitor impacts of climate change
- Reducing consumption in line with planetary limits
- Universal respect of Mother Earth rights as captured at the Cochabamba Peoples Summit on Climate Change.

- Leave the fossils in the soil. Besides global warming, the environmental cost of fossils cannot justify a continued reliance on the resource. Reflect on Shell's pollution of Ogoni land as captured by UNEP. Think also about the open scars created by tar sand extraction in Alberta, Canada. Think about Texaco's destruction of the Ecuadorian Amazonia. Who benefits from all that? Certainly not the planet!
- Set up Climate Tribunals to try Climate Criminals – the unrepentant polluters whether heads of corporations or states. Ecocide is no less a crime than genocide.

CONCLUSION

Our narrative must be the story of our lives told by us and dipped in our experiences.

> ... If there is any hope for the world at all, it does not live in climate change conference rooms or in cities with tall buildings. It lives low to the ground, with its arms around the people who go to battle every day to protect their forests, their mountains and their rivers because they know that the forests, the mountains and the rivers protect them. The first step toward re-imagining a world gone terribly wrong would be to stop the annihilation of those who have a different imagination— an imagination that is outside capitalism as well as communism. An imagination which has an altogether different understanding of what constitutes happiness and fulfilment.[11]

It is our life, we know how the rain has beaten us and for how long. Our narrative must not be stuck in the crisis narrative imagined about us by others. We must awake, arise, mobilise and work for the transformation of our society and planet— by all legitimate means available and necessary.

11. Arundhati Roy (February 19, 2013): 'Decolonize the Consumerist Wasteland: Re-imagining a World Beyond Capitalism and Communism,' *CommonDreams*, http://www.commondreams.org/views/2013/02/19/decolonize-consumerist-wasteland-re-imagining-world-beyond-capitalism-and-communism . An excerpt from her book, *Walking with the Comrades*.

PART II

VIOLENCE IN THE LAND

14

As Kogi fights over refinery location

This article was first published July 2010 in 234NEXT[1]
The struggle over the location of a refinery in Kogi State has caught the attention of many Nigerians. The governor is accused of taking the refinery away from Lokoja to his hometown.

The submission of this article is that the governor, and all those who made the deal with the Chinese to build three refineries, should actually be forced to locate these refineries in not just their villages, but on their own private land as well.

Why?

Refineries are not industrial installations that people should wish to be located even in their enemy's community. They are extremely toxic and poison everything and everyone around them. This is well known in the communities close to refineries in Warri, Kaduna, and Port Harcourt.

Apart from the release of toxic gaseous emissions into the atmosphere, the liquid effluents from these refineries are scarcely treated, and are dumped into water bodies on which local communities depend. The case of Ubeji community, behind the Warri Refinery, is particularly pathetic.

The community's river and their mangrove swamps were severely polluted and engulfed in flames in July 2007. Till date, no remediation exercise has been carried out. You may hear that some compensation has been paid, but what is that pittance compared to the danger to which the community is permanently exposed to? What would such minor compensations do when the livelihoods of most of the citizens have been more or less permanently curtailed?

Other countries' examples

The toxic impacts of refineries are just as bad in other parts of the world. In South Durban, South Africa, the refineries (owned by Shell/BP joint venture) were located according to the dictates of the apartheid political system.

A visit to these communities today reveals a high incidence of cancers, blood disorders, and respiratory diseases such as asthma. Indeed, the prevalence

---

1. The article was also published as Kogi Fights Over Refinery Location at http://nigeriang.com/money/as-kogi-fights-over-refinery-location/3151/ (accessed 31 May 2016)

of cancers and asthma is so high that you would hardly find a family without members who have died from these diseases, or who are suffering from them. One of the things kids pack as they head to school is the pumps to use in suppressing asthmatic attacks.

The difference between the refineries of South Africa and the ones in Nigeria is that the communities there are organised against pollution and work to produce evidence through the use of means such as the Bucket Brigades (who use bucket-like equipment to collect air samples for measurements).

There have been charges of environmental racism with regard to the location of toxic factories in the USA. One of the most spectacular incidents involving a refinery in the USA was the huge explosion that occurred at the Shell refinery at Norco, Louisiana, in May 1988. The fire from that explosion lasted for eight hours before it was contained. The blame was placed on rusty pipelines and inadequate preventive maintenance procedures.

There are several examples around the world of the negative consequences of setting refineries in neighbouring communities. One peculiar case is an aged Shell refinery in Curacao (near Venezuela) now being run by the Venezuelan state oil company, after Shell sold the refinery to the Curacao government in the 1980s for less than one dollar. They sold the refinery because they were faced with the need to clean up toxic dumps they had created at a cost of about 400 million dollars.

Back to Nigeria, it is mind-boggling to find people fighting to have these installations in their localities. Those whose localities they are moved away from should actually be engaged in thanksgiving and celebrations, rather than blocking highways in protests! The Chinese have found a business opportunity because the NNPC has been inept at managing the four refineries in Nigeria. Must the need to meet increasing demand for petroleum products force us to open ourselves to be ripped off?

The Chinese are to build and run the refineries until they recover their investments. Without terminal dates of when CSCEC would hand over the facilities to the NNPC, there is a wide room for corrupt practices and unmitigated exploitation.

Moreover, placing the refineries on the banks of the River Niger in Kogi State, as well as on the shores of the Atlantic at Lekki may be ways of democratising pollution, but these are moves we can ill afford at this time.

Besides, we need public debates and examination of environmental impact assessments for these projects before they proceed further.

15

Violence in the land

This article was first published in 234NEXT on January 6, 2011[1]
Things have a quirky way of becoming the vogue in Nigeria. And once entrenched, unlike fads that come and go, these do not easily fade away. Think of the funny emails often written in all capital letters and in very bad language. They make many people laugh. But they also trap many others who are as greedy as the fabricators of those mails. I cannot say if such 419 soliciting started in Nigeria or if our compatriots simply caught up with it and took over the trade. Whatever is the case, upper case e-mail scams now have the reputation of being mainly a Nigerian phenomenon.

Nigerians did not invent the business of kidnapping. However, once it left the realms of tales and took concrete foothold in the Niger Delta, it became a Nigerian nightmare. In places like Aba, financial institutions had to close for some time because of the spate of kidnappings and general insecurity. You would think that only the rich got targeted. No. Being rich or poor makes no difference to the predators beyond the size of the cash they could extort from the related families, associates, corporations or government. Politicians, oil workers, journalists, business people, the clergy, school children and just about anyone became fair game.

When we examine the trend closely it does appear that the manifestation of the levels of primitive violence on our shores can be linked to fraud. In other words, what we may well be witnessing is a manifestation of fraud in its most crude form. And fraud appears to be very lucrative here because even when caught, the punishment is a slap on the wrist.

When kidnapping kingpins saw that taking oil company workers hostage was a quick way of latching on the national looting train, they dug in and extended their networks. When others saw that they could get their parents or relatives to part with cash, they arranged to get 'kidnapped' and by that broke through to their supposedly selfish folks. Relatives entrusted with the care of children suddenly became kidnappers and others sent to pick up children from school suddenly developed wings and orchestrated the now well-worn trade. Who would say that this is not a manifestation of the 419 bent? If corporations,

---

1. (no longer accessible)

governments, and security agents had refused to play ball right from the onset of this phenomenon would it had grown to the current proportions?

The current fad is to drop bombs with intent to wreak havoc on life and property. The origin of this sort of violence is not Nigerian. There are certain countries and regions that have been wrecked by this sort of senseless destruction for decades now. Today, Nigeria risks becoming one of such nations. Here, festive seasons have become preferred times to kill people physically, and also to unleash social violence in the resultant ripples. And so we witnessed the bombings in Abuja on national independence day. While the military brass band struck marching notes, the harbingers of death triggered their bombs. And on Christmas Eve, bombs went off in Jos claiming innocent lives. The incidents in the Maiduguri area are almost becoming routine. On New Year's Eve, while other nations ushered in the second decade of the millennium with artistically engineered fireworks, the agents of destruction set off bombs in a military barrack in Abuja.

Poverty fuels violence

The violence is promoted by certain factors. One is the entrenched poverty. This poverty has both financial and mental dimensions. Mental poverty promotes votes rigging and other forms of electoral fraud. Politicians who are used to getting into office or positions through fraudulent processes use violence as a vital tool for achieving their aims. An example is the mindless killings in Ibadan during a local government congress of the People's Democratic Party. The same can be said of the bombings at a political rally in Yenagoa, Bayelsa State. In Akwa Ibom State there has been a trend where a declaration of intention to run for certain political offices has meant an invitation to violent reactions on such individuals or their next of kin.

What will happen as party primaries begin and as the election days arrive? Will it be safe to drop a ballot in the box without the box exploding beneath our hands? It is sad that at a time like this, some politicians would misapply a well-known political statement that now positions them as supporters of violent change.

Is there a chance that a nation exposed to this level of primordial violence can get out of it without long-term scars? It will amount to wishful thinking for anyone to assume that the violence in the land would not have lasting effects on our national psyche. The violence has pushed the notion that it is dangerous to engage in honest labour and that you need to be a purveyor of violence before you can be a factor to be reckoned with in the political scheme of things.

It is a known fact that environmental factors such as entrenched pollution, as well as drastic social events, affect not just the generations who witness such events but also those that follow. These shock waves may impact the genetic information passed on to future generations at all levels. When major shifts occur in quick successions, the disorienting effect can be massive. Just imagine a cultural shift occurring within a generation. We are experiencing this in Nigeria although some may claim this to be a global phenomenon.

Doing the right thing has suddenly become obnoxious. Fraud is celebrated and rewarded and often times with chieftaincy titles. Where did all these start and where would they end? Bob Marley's suggestion (in his song, Real Situation) that total destruction may be the only solution is anarchistic and we do not recommend that. But will we continue to accept fraud and violence as the norm or shall we get angry enough to trigger organised resistance?

16

A nation split by oil

This article was published on 13 January 2011[1]

As Sudanese vote this week on staying as one nation or becoming two, my mind goes back to when civil war broke out in Nigeria in 1967. I recall that when Biafra was announced, I leapt in celebration at the novelty of suddenly being a citizen of a new country under a new flag and with a bearded man at the head of state. What my young mind could not fathom, and did not question, were the reasons for the emergence of the new nation. What were the announced reasons and what were the unspoken ones?

Before we could settle to savour the change expected from the split, things took a different turn. The war drums sounded, and bullets began to fly. Streams of refugees flooded through our village and soon enough, we were on the move. I still recall seeing starving kids, rotting corpses by the roadside, and I can hear the screams of young ladies who were captured and forcibly married by rampaging troops.

We see the great mobilisations by the peoples of Southern Sudan for a split and when the result of the referendum is announced, we can bet that the result is like a dream long foretold.

There are many reasons why the South should be eager to drift away. Indices of development from the country are severely skewed against the region. Reports have it that over 80 per cent of the inhabitants of Southern Sudan have no sanitation facilities.

While almost 70 per cent of the people living in Khartoum, River Nile, and Gezira states have access to pipe borne water, the people in the south depend on boreholes and rudimentary water wells. They and those in the Darfur area depend largely on food aid for survival on account of the dislocation of the agricultural sector by entrenched violent conflict.

Certainly, all will agree that oil is a major factor in the political fortunes of Nigeria. We may squabble and bicker under the cover of ethnic or regional differences, but beneath the surface, the struggle is over who controls the massive oil and gas resources and revenues of the land. The struggle for power

---

1. Sudan and oil politics: A nation split by oil. http://www.pambazuka.net/en/category.php/features/70056 (accessed 31 May 2016)

at the centre was set the moment a unitary system of government was decreed in 1966 and has since coloured the sort of federal system that the nation runs on.

Oil is a principal factor in the current political situation in Sudan. Exploration activities started in the 1960s by AGIP, the Italian oil company, which found natural gas in the Red Sea. The American oil giant, Chevron, followed suit but never revealed what they found, according to reports.

Like Nigeria, like Sudan

As time went on, a number of Chinese and Asian companies jumped in and finally oil was produced from the Muglad Oil Basin, Blocks 2 and 4. Sudan is divided into 17 oil concession blocks with SUDAPET, the government owned company, working in joint partnership with the various Asian and European oil companies.

As aptly captured by a Sudanese academic in a recent Oilwatch Africa meeting, 'Sudanese oil has been developed against the background of war, international sanctions, and political isolation. It has been developed at a time of imposing demand by emerging economies like India and China and a time of unprecedented soaring prices of both food and oil and the controversial use of agricultural crops as a source of bio-energy.'

Quite like Nigeria, oil produces over 75 per cent of the foreign exchange earnings of Sudan. Other production sectors have equally been almost completely neglected. Before oil, over 50 per cent of Sudan's revenues came from the agriculture sector, contributed 95 per cent of the export earnings, and employed a high percentage of the total labour force in the country.

With oil as a major economic factor, and seeing that the bulk comes from the South, developments nevertheless eluded the region. An example can be seen in the first refinery which was sited about 70 Km north of Khartoum. Crude export pipelines run northward and amount to about 5326 km in length.

The reality is that with the available infrastructure, the South cannot export its oil except through the North. In addition, as the date of possible separation drew nearer, new oil blocks that transverse northern and southern areas were being allocated.

Oil companies operating in Sudan are exempted from paying taxes. The contracts were mostly negotiated when the price of an oil barrel of oil was less than 20 US dollars. Surely, the companies operating here could not hope for a better space for reckless exploitation and incredibly high profit margins. Added to this is the fact that the regulatory regime is largely non-existent and even the conduct of environmental impact assessments are selective.

With Sudan having about five billion barrels of oil in reserves and currently exporting billions of dollars' worth of oil per year, it must be painful for Khartoum to let the oil-rich South go. About 80 per cent of Sudan's oil exports come from the southern states. Only 50 per cent of revenue accruing from oil goes to the South, a factor that undoubtedly stokes the embers of discontent in the area.

As the peoples of Sudan vote for the emergence of a new Southern nation, dreams of the desperately poor and those traumatised by war and cruelties will run high. Children who never experienced peaceful environments will be marvelling at great possibilities. Oil has certainly greased the engines of exploitation, oppression and war in Sudan. It is oiling the machines of separation today. What will it lubricate next?

These are questions we must mull over, but a bigger question is over the implication of continued fragmentation for Africa as a whole. At a time when the continent should be coming together and erasing the arbitrary boundary lines drawn by colonialist adventurers, we continue to fragment. Certainly, this cannot be the only way to overcome poor and parasitic governance.

17

The 'milking' of oil workers

This article was published on my blog, May 2013[1]

If you have ever passed by the entrance gates to any of the oil companies, you must have seen a warning sign that says that you would not be allowed to drive into the premises without using your seat belt. The intent of those signs is to indicate to you that the companies follow strict safety rules. Some years back, I was in a training meeting of directors of the Nigerian National Petroleum Company (NNPC). The major focus was on how to handle rampant pipeline problems. At that meeting, it was revealed that in trying to clamp ruptured or damaged pipes conveying refined petroleum products, some workers had to stand in pools of petrol or diesel to carry out their assignments, obviously without adequate protection.

Generally, oil field workers are as exposed as communities are to the dangerous pollutants of the industry. At that meeting, I had an opportunity to propose the thesis that 'sabotage' must be seen in some contexts as a legitimate political weapon. Legitimate? While remaining a proponent of non-violent resistance, it must be recognised that unless sabotage is seen as a possible weapon for the expression of dissent, then the right solution to the problems may never be found. The thesis was roundly rejected. But eventually, when the sparks started to fly in the oil fields and in the surrounding communities, it began to sink in that the ultimate solution to address the explosive dissent in the Niger Delta must be found in tackling the root causes of the dissent.

A couple of weeks ago, I was opportune to participate in health and safety workshops for oil sector workers, organised by the American Center for International Labor Solidarity, also known as the Solidarity Center. There were four workshops in all, but I was only able to attend the ones in Warri and Lagos. While one cannot compare working in the oil fields to working in violent conflict zones, or at a nuclear power plant, it is quite true that workers in the oil and gas sector need to be pretty much concerned about health and safety issues. Some of the workers who perform sedentary duties in offices complained that the constant focus on computer screens poses serious health issues to them.

Others said that they were required to ensure they grab staircase handrails

1. http://nnimmo.blogspot.com.ng/2013/05/the-milking-of-oil-workers.html (accessed 31 May 2016)

while climbing or descending the stairs to avoid falls. That must be why we have those banisters and balustrades, surely? A machine operator complained that he has hearing problems due to exposure to extreme noise at his workplace. Some of the field workers said that sometimes they have to climb dangerous heights while performing their duties on the rigs and other locations. Even though they wear safety belts, the dangers are always there.

Management versus workers

There were healthy debates over the question of who was to be blamed for most workplace accidents in this sector— management or workers? It was striking to see oil-worker unionists speaking almost like their managers. Some informed the workshops that they did not have safety issues in their companies because management took care of everything. Quite a number of them believed that the management did the utmost in providing safety gears and took other measures that should keep accidents from happening. They maintained that the blame must be placed on the workers since it was likely the accidents took place when workers cut corners or otherwise ignored specified methods and processes.

Those who held that the management were to be blamed for most of the accidents insisted that management cared more about machines than they care about the workers. They held also that accidents do happen even when the procedures set out by management are followed. Another point was that, sometimes, workers are forced by management to take shortcuts in order to meet production targets.

One interesting fact shared at the workshops was the need for unions to carry out workplace mapping of health and accident issues. When such mapping is done over time, a pattern of accident or health issues related to particular workstations or procedures emerge. It was also noted that the shop floor workers could take these steps, even if the unions are not keen on monitoring and mapping.

Milk as antidote

A real surprise that came from these workshops was the revelation that oil companies use milk as an antidote for exposure to heat and hazardous chemicals. We exchanged banter that if cow milk was so efficacious and could cure cancers and other health challenges, then every oil worker should own a milk cow. A participant from one of the top oil transnational corporations said that the company provides tins of milk to workers who man their electricity generators to counter the impact of the heat and chemical exposures. It sounded as a joke initially, but it turned out to be a serious matter. A former union member said that between 1975 and 1978, while working in a gas industry, the workers, who produced acetylene and oxygen in cylinders, were always provided with milk while on duty. Medical experts will have to tell us if milk is the antidote to heat and chemical exposure in the oil and gas sector. Could this be another way by which workers are taken for a ride, exposed to harm, and then given a false sense of well-being through gifts of tins of milk? If this is fraud, the companies who engage in this deception must be brought to book. If it is effective, then get me my cow.

18

The tragedy of Ayakoromo

*This piece responded to the heavy-handed assault by Nigerian military on a
Ayakoromo community in the Niger Delta in May 2011*[1]
It is difficult to resist the temptation of writing about the unfolding turmoil
in the Maghreb region. The events in Tunisia, Egypt, and other countries that
up until recently seemed untouchable by popular revolt is instructive in many
ways.

In sub-Saharan Africa, we appear to be sitting on the ringside, buffered by
the dessert, maintaining largely deserted streets, and probably see the uprisings
up north as opera.

The courageous uprisings in North Africa reveal the complexity of history.
Just take a look at Cote D'Ivoire with two presidents and divided streets. Cabals
with political leverage have played various cards to maintain their hold on
power and as long as the people are divided, their reign is secure.

It is also instructive that a movement can erupt without a physical icon or
individual leader and led by even loose collectives as we see in Egypt.

But this piece is not about all of these places. It is about our own backyard,
Ayakoromo in Delta State. This community received an end of year package
from the Nigerian military on December 1, 2010 when bombs and other
weapons of war were unleashed on it. Their crime? There was, or had been, a
militants' camp in or near the community.

According to reports, the Joint Military Task Force (JTF) attacked the
community in their effort to apprehend or annihilate John Togo, the leader of
a group known as the Niger Delta Liberation Force (NDLF). When the news
broke, the JTF announced that they had captured and destroyed the camp of
the NDLF. However, the militant group claimed that they had destroyed the
camp themselves and relocated weeks before the attack.

The attack resulted in extensive destruction of property and displaced
thousands of innocent folks who had to run to refugee camps in Warri and
environ. The aged, infirm, and others who survived the raid but could not run
away apparently remained in the community which was taken by the troops.

---

1. http://nigeriang.com/money/oil-politics-the-tragedy-of-ayakoromo/7349/ (accessed 31 May 2016)

The number of lives lost is contested. The community has used stakes to outline a spot they claim is a mass grave of the victims. This is a totem to an outrage.

Dreams betrayed

The tragedy of Ayakoromo is the tragic manifestation of dreams betrayed in the evolution of our national history. Ayakoromo underscores the fact that the citizenry of this nation have not, in any deep qualitative way, enjoyed better respect of their human rights under military autocracy or under democratic structures. Apart from the casualties of the civil war, more lives have arguably been lost under civilian rule than under the military.

We are in no way nostalgic about the days of the jackboot, but think of what the ordinary people have suffered since 'agbada' replaced 'khaki' in the corridors of power.

Odi happened soon after the return to civil rule in November 1999 when the town was shelled (no metaphor meant), bombed, and wrecked by the Nigerian military on the pretext that they were searching for some 'kidnappers'.

In the attack, about 2,800 lives were wasted and a blanket of silence[2] still shrouds that monstrous assault. There has been no inquiry, and those who survived learnt their lessons from the several graffiti left behind by the rampaging troops who were obviously out on a mission to decimate the local population.

This was followed by the attack on Odioma.[3] Last May, the Gbaramatu kingdom of Delta State received a dose of the lethal medicine.[4] We are not mentioning several cases of lesser magnitude that have occurred in-between.

Consider also the unravelling events in Jos, Maiduguri, and Bauchi.[5] Bombs are used freely and now lynch mobs appear to have stepped into the fray. What do these portend for the forthcoming elections? The mass response of the citizens to get registered may be an indication that Nigerians are ready for change, to do things right. Are our leaders ready? One can only hope that we are not waiting until folks immolate themselves and trigger a Tunisian run.

Every assault is treated as being of no consequence. There are no enquiries. There are no punishments. Sometimes, there may be grudgingly given

2. Environmental Rights Action/Friends of the Earth Nigeria documented 2483 names of victims of the Odi massacre. See A Blanket of Silence by ERA/FoEN at http://www.eraction.org/publications/silence.pdfaccessed 31 May 2016

3. The attack on Odioma, Brass Local Government Area of Bayelsa State, Nigeria, took place in February 2005. More than 100 persons were killed in that incident.

4. Several communities in Gbaramtu Kingdom of Delta State, Nigeria, were bombed by the Nigerian Military's Joint Task Force (JTF) in May 2009. The stated reason was that they were chasing after Niger Delta militants that had camps in the area. The number of casualties are not precise. Sweet Crude reported 500-2,000 deaths. http://www.sweetcrudemovie.com/attacks.php (accessed 31 may 2016). Thousands were displaced and spent months in refugee camps before they could return home to pick up the pieces.

5. These were the early days of the violent reign that came to be know as Boko Haram insurgency

apologies, but generally, justice is not served. The streets of Jos and creeks of the oil fields run with the blood of the innocent. Is the life of the poor of such little value that we can simply shut our eyes and move on as though nothing has happened?

Certainly, we cannot afford a reign of terror either from the military or armed gangs in our country. The onus lies on the government to provide security for the Nigerian people. The Nigerian security forces cannot be allowed to terrorise, kill, and destroy at will under any guise. Where individuals run afoul of the law, it is the job of law enforcement agents to fish such out and bring them to justice through constitutional avenues.

The time has come for the books to be opened and all the cases that have been swept under the carpet openly examined in a special commission of enquiry. Offenders, military or civilian, should be appropriately sanctioned. Other elements of justice must encompass restitution; including the rebuilding and upgrading of destroyed communities. The tragedy of Ayakoromo must not be repeated.

Donations of blankets and rice to victims of these attacks may be good, but the real relief will only come when governments own up to their responsibility to protect lives, apologise to the people, and commit never to turn out troops against the people.

We see pictures of jet planes and helicopters flying low over protesters in Egypt. We see protesters step on armoured tanks. Here, when helicopters, gunboats, and air force planes swooped over Gbaramatu and Ayakoromo it was not to warn anybody. It was to bomb, level, and kill. This must stop.

19

Mending MEND

This paper was published on October 07, 2010 in 234NEXT[1]
Nigerians have been subjected to several years of autocracy, misrule, and serial abuses these past 50 years of flag independence. The Movement for the Emancipation of the Niger Delta (MEND) and other groups has said that Nigeria has no reason to mark this 'jubilee'.

MEND did not only make the point that there should be no celebration, they went ahead and set off bombs that snuffed the lives of over a dozen Nigerians and maimed many others.[2] That was certainly a strong way to make a point— in broken bodies, spilled blood, shattered families, and stunning the nation to boot.

People have reacted in different ways to the Abuja bombings, a remarkable escalation of the sense of insecurity in a nation where kidnapers do not care a hoot about taking kids, journalists, pastors, oil workers, and just about anyone into captivity. This is a nation where citizens are abandoning their homes, villages, and towns for armed groups to take control and turn them into camps for their 'armed struggles'. Meanwhile, the security organs are out on road blocks asking 'wetin you carry?'

The idea of not marking national days in the country crept into the national psyche from the years of military misrule when the dictators did not wish to promote the assembly of peoples to discuss the national state of affairs. It became fashionable to tell Nigerians that occasions such as independence anniversaries, children's day celebrations, and others were moments for sober reflection.

This was actually a way of camouflaging the fact that the leaders were utterly bereft of any ability or inclination to reflect on much other than their piles of loot. Over the years, this neglect became accepted as times to stay in our homes, mourn the death of dreams built on the 'labours of our heroes past' that are now threatened to have been in vain.

By neglecting to mark days such as that of national independence, the

---

1. No longer accessible online

2. Nossiter, Adam. (1 October 2010), Bombs by Nigerian Insurgents Kill 8.
http://www.nytimes.com/2010/10/02/world/africa/02nigeria.html (accessed 31 May 2016)

remaining threads that give citizens a sense of nationhood kept being pulled out of our multi-coloured national social fabric. Soon, we consolidated our sense of apartness, each looking more to our ethnic nations, regional cleavages, and political cabals.

It is in that trajectory that we read the unfortunate order from MEND that no one was to go to the Eagle Square for the national day celebration. They were kind enough to say that people should avoid dustbins and cars. Pray, where were those who eat out of dustbins going to get their meals from? Or had MEND dropped extra packages for them to gather?

Of all the responses, the one that is perhaps the most poignant is that of President Goodluck Jonathan. In the chorus of voices condemning the assault on all of us, our president reportedly said 'What happened yesterday was a terrorist act and MEND was just used as a straw; MEND is not a terrorist group.'[3]

By his leadership position, Mr. President certainly has more information on security matters than us ordinary citizens. Two disturbing issues arise from his assertion. The first is his conclusion that 'MEND was just used as a straw.' The first assertion is more alarming than the second one which claims 'MEND is not a terrorist group.'

Perhaps, MEND is a political party or an extension of the Nigerian Army, Mr. President? Or is this an exercise in socio-political engineering to mend MEND?

Straw or pawn?

We return to the first assertion, which suggests that MEND is naive and lent itself to be used as a straw. In trying to read the president's lips, we assume that he was using the word straw here to mean 'pawn', referring to someone used or manipulated to further someone else's purposes.

If MEND is being used to further the purposes of someone else, then we have reasons to raise more concerns.

The first is that that someone has to be unveiled. Another concern would be to fundamentally question the rise of armed groups in the Niger Delta allegedly fighting for a number of things, including more oil and gas revenues for the region. Have there always been puppeteers behind the scene if the armed groups do not have agenda for their activities? This is disturbing because many came to see MEND as one of the more politically coherent groups that chose the way of violence to make their points.

If MEND is a straw, can we assume that scenario planners, who have predicted that Nigeria will blow into pieces within a short space of time, have an interest in the escalation of violence and insecurity in Nigeria? Are we to say that the violence in the oil fields has not secured sufficient foothold for foreign armed assistance and this needs to be extended to the entire nation and

3. This was a rather cryptic claim as the President did not explain what made a group 'terrorist' or not. http://www.informationng.com/2010/10/bomb-blasts-terrorists-not-mend-responsible-jonathan.htmlaccessed 31 May 2016

possibly put Nigeria on the path to becoming another Somalia or even Sudan to a degree?

If MEND is a straw, at what point did they metamorphose into this, or were they straws right from start? If the group is a straw or can be used as a straw, what are/were the several others who embraced the amnesty programme of the government? It is time to rethink the amnesty programme and extend it to the damaged environment of the region and indeed of the nation through a national environmental emergency plan.

The president's assertion requires serious interrogation. With the background that some armed groups began as bands of political thugs, we need to know if this assault on poor Nigerians is linked to the fight for space and displacements in the run for the forthcoming elections. In other words, were these explosions the hands of politicians but the voice of MEND?

What we have here is a deep failure of our security systems and this requires quick action by Mr. President, and not quaint definitions of what constitutes terrorism.

20

The amnesty worked

This was written in December 2010 in reaction to the success of the amnesty extended to Niger Delta Militants. It was published in 234NEXT newspaper as well as on my blog[1]

The rise of crude oil price in the market raises hope of boom time for producers of the resource and fears of high-energy costs for others. Price thresholds above US$80 per barrel also make investment in some forms of energy such as agrofuels appear attractive. For Nigeria, as the price of crude inches up, so must the gobblers of so-called excess crude funds be getting ready for the kill.

As the major supplier of government revenue, the crude oil price rise must be accompanied by an increase in production to ensure maximum benefit to the government and the oil corporations. This would mean keeping all oil wells pumping at full throttle. It would also mean ensuring that peace reigns in the oil fields, even if it means exerting maximum firepower in search of a handful of renegade post-amnesty militants.

The popular spaces in Cancún began to fill up over the last weekend, even as the climate talks got ready for the home stretch. The environmental justice movement believes rightly that fossil fuels must be left in the ground, as their use is responsible for the release of much of greenhouse gases in the atmosphere. Leaving the fossil fuels in the soil would translate less pollution and less toxic compounds in the environment. It would also mean rapidly transiting to renewable or less harmful energy sources and into a post carbon civilisation.

Negotiators in the climate talks are not listening to the clarion call to leave the fossil fuels in the soil. What is music to their ears, however, is how the carbon that is released when the fossil fuels are used can be captured and stored. No, they are not exactly debating the best technologies that can achieve this. So, what is on the table?

Climate negotiators are seeking to make carbon capture and storage

1. http://nnimmo.blogspot.com.ng/2010/12/amnesty-worked.html?q=OIL POLITICS: The Amnesty worked: Over the years, conflicts have been orchestrated in the Niger Delta. (accessed 31 May 2016). I also mulled over this issue also in my book, *To Cook A Continent: Destructive Extraction and the Climate Crisis in Africa* (Pambazuka Press, Oxford, 2012)

projects eligible for carbon credits. Technologies for capturing and storing carbon are far from being ready for implementation at the moment. There are also issues over costs as well as doubts over their effectiveness. However, leaving the fossil fuels in the soil is undoubtedly effective carbon capture and storage. This option does not require technology transfer. Neither does it require any capital outlay.

Challenging the reckless nature of the oil industry, I was privileged to join a team of nature defenders to institute a case in the constitutional court of Ecuador against BP for their reckless activities and oil spill in the Gulf of Mexico.[2] The case opens a unique way for holding corporations and individuals accountable for their acts anywhere in the world. It is also a direct action in tackling climate change. Two of the key demands of the case is that BP should leave as much oil as they have spilled in the ground and should stop deep water activities.

Will the world's addiction to crude oil allow the voice of reason to prevail? Will the climate negotiators pause to review all the false solutions plastered on the negotiating texts by corporate interests fuelled by greed as well as the creed that the market holds the solution to every problem?

Assaults in the creeks

While the price of crude oil increases and yields more revenue to both the government and the oil companies, the environmental and social impacts are still externalised to the poor communities. To ensure that oil must flow at all costs, it does not appear to matter how much human bloodletting happens in the process.

Over the years, conflicts have been orchestrated in the Niger Delta— and indeed other parts of Nigeria— either for economic reasons or for political ones. When the late President Yar'Adua announced an amnesty for the armed groups in the oil fields, popularly known as militants, critics doubted that the amnesty would work. Others simply prayed that it would work. And it did.

The amnesty programme had some foundational problems because of the nature of the conflicts on the ground. Usually, combats involve taking of territories or for political supremacy. The fights in the Niger Delta is not one for territorial control, neither is it for political power. It can be, and has been interpreted, as largely opportunistic and as means for capital accumulation.

However, it must again be stated that some sense of political disenchantment is also discernible. In all the expressions, the environment continues to suffer; the local communities continue to be carpeted through ground, sea, and air bombardments.

We remember what happened to Gbaramatu Kingdom in May 2009. After

2. Democracy Now! November 29, 2010. BP Sued in Ecuadorian Court for Violating Rights of Nature. http://www.democracynow.org/2010/11/29/headlines/ bp_sued_in_ecuadorian_court_for_violating_rights_of_nature (accessed 31 May 2016)

the assault, 3,000 women with their kids became refugees for months at a health facility in Ogbe Ijoh. Now, with the latest levelling of Ayakoromo community, Delta State, the same health facility has again become home for displaced local people. That health facility is a clear metaphor for the jaundiced development efforts in the region. If it were functioning as a hospital, as it was designed, would it readily turn into a refugee camp?

The resumption of open hostilities says something about the amnesty programme. That scheme was built on mostly accumulated military hardware and personnel in the Niger Delta, and spending a tiny fraction of the overall budget on training and reintegration of repentant militants.

Reports have shown that many youths who requested to be trained and rehabilitated could not be taken on because of some quota system that had already established a ceiling as to how many could be trained. According to Dutch media reports, companies such as Shell have hired some of the retrained militants as welders and fitters. That also tells a story on its own.

But the real issue of deep environmental pollution is yet to be tackled and unless the environment is safe for local people to return to their normal means of livelihood, any declared amnesty is a smokescreen and is bound to blow up in smoke. However, when all is considered, we can submit that the current amnesty has worked beyond what it was designed to achieve.

PART III

EXTRACTIVES AND TRANSPARENCY

21

When oil companies volunteer

This article first was written for 234NEXT in November 2010[1]
Since oil companies gained dominance of the world economic system, literally driving the engines of industrialisation and modern fossil civilisation, they have taken several steps that have endangered humanity.

The massive burning of fossil fuels, such as oil and gas, has contributed immensely to the stoking of the atmosphere with greenhouse gases responsible for global warming.

The sector is also known to have been responsible for environmental and human rights abuses in the world. The presentation of their commodity as the cheapest form of available energy has been sustained over a century by cost externalisation to the voiceless, whose environments have been heavily assaulted. The energy wars that are sometimes masked as war on terror are also well known. The contribution of oil companies to human misery is well documented.

Although the leopard may not change its spots, the companies have not been blind to the woes they generate. One of the steps they have taken to cushion the impact of their harm has unfortunately been nothing more than hogwash. One subtle way this has been done has been to plant into public minds that they are not oil, but energy companies. The difference may be subtle, but it seeks to erode the stink that the former name carries. We insist on calling them by the name that best describes them and to avoid grouping them in the same slot as clean energy producing companies.

Apart from change of nomenclature, the fossil fuel sector has etched some oxymorons into public minds, making people accept clearly contradictory terms as being logical. Take the example of clean coal. What is that? There are others, but this is not the focus of our discussion today.

Voluntary Principles on Security and Human Rights
Some oil companies, including Shell and Chevron, have signed up to what

---

1. It can be accessed at http://nigeriang.com/money/oil-politics-when-oil-companies-volunteer/5350/ (accessed 7 june 2016)

is known as Voluntary Principles[2], by which they solemnly declare how they would change their corporate practices in the area of security and human rights.

The question this raises is whether the endorsement of these voluntary and non-binding principles has brought about any positive change. The Voluntary Principles are not even known to be in existence by many. We will touch briefly on some key areas of the principles. You are urged to ask how those principles are applied in Nigerian oil fields.

The companies say they will report payments made to security forces or, in our case, to the Nigerian government for supply of security cover for company operations. If such records were properly kept, it would be possible for such companies to be held accountable where funds are tied to incidents that resulted in human rights abuses. If a company pays money to the military, for example, and the funds support an assault on a community, the link should be transparently traceable for this clause to make sense.

A look at the Voluntary Principles appears to start from the premise that oil company security depends on the actions of the country's security forces. This thinking has maintained the relationship with the Nigerian military and police and continues to encourage abuse. It also often precipitates clear acts of mayhem. Oil companies sometimes review their security arrangements to determine if the relationship they have built with the security forces has been a credit or a liability.

A review conducted by Chevron in 1999 found that Nigerian security forces were actually more of a liability than a benefit[3], and that they were prone to cause great harm both to Delta residents and company employees. Shell, on its part admitted in a 2003 security review that it had contributed to the rise of conflict and corruption in the Delta region through its relationship with security forces. The question is, what changes have they made?

We submit here that if the official security forces provide a safe atmosphere for ordinary citizens, corporate citizens would also enjoy the same. Moreover, if oil companies maintain their equipment, operate with the same standards they apply in their home countries, and respect community rights, there would be no need for special security arrangements that must be eating into their resources.

The Voluntary Principles also require that oil companies communicate effectively on Human Rights Principles to security forces and ensure proper training, and screening of known human rights abusers.

Security officers of corporations and public security forces are often tied

2. See the principles at http://www.voluntaryprinciples.org/ (accessed 7 June 2016)

3. This was stated in theDeclaration of Scott Davis in *Bowoto v. Chevron Corp.*, para. 41 (filed in the USA, Nov. 22, 2006). See also at https://ccrjustice.org/home/get-involved/tools-resources/fact-sheets-and-faqs/factsheet-case-against-shell where we are told that in the Bowoto case, 'The victims of the [armed forces] were targeted because they were oil protesters, or because they were associated with oil protesters . . . . Though the evidence indicates that the [armed forces] were not particularly selective in choosing their targets, the victims . . . were not targeted . . . simply because they were civilians.' P.22 Weblink (accessed 7 June 2016)

together in mutually dependent arrangements, whereby governments take primary responsibility for security and the private entity provides resources and logistical support. To what extent have the guidelines provided in the Principles been used to ensure that the conduct of the forces abides by human rights law?

Holding Individuals Accountable

It is known that oil companies do keep security logs showing records of security incidents as they occur at their facilities. They should also be required to keep full records of incidents in which local residents are injured or killed in confrontations with government security forces, acting to secure the interest of the companies. Such incidents should also be reported promptly and publicly. Individuals indicted should be held accountable.

The Voluntary Principles provide an opportunity for the Nigerian legislative houses at the state and federal levels to take their provisions, review, and enact them into law. The oil companies may have endorsed the principles as a way of beefing up their public image and presenting the face of companies that care about human rights.

Enacting same into law will encourage the companies to implement them by making them mandatory principles. It will also help the companies to bridge a part of the huge deficits they have accumulated in terms of transparency in their activities.

22

Environmental issues in extractive industries transparency

This paper was first published in 2012[1]

Reckless crude oil extraction is violence visited on the environment and people today and an unconscionable draining of the wealth of future generations.[2] Transparency in the extractive sector is not complete without taking cognisance of the impacts operators and operations have on the environment. Transparency issues must include measures of the extent to which operators adhere to environmental and industry standards. As this chapter focuses on the oil fields of the Niger Delta, we declare that unless an audit of the environmental degradation of the region is carried out and a master plan for the detoxification of the environment is designed and implemented, talk of transparency in the sector will remain at best superficial.

Environmental rights must be seen as the human rights that they are. They are also the most commonly trampled upon rights and this results in a loss of ability of the people to enjoy any reasonable level of wellbeing in their land. The destruction of the Niger Delta environment calls for emergency-scale restoration actions to stem the tide of loss of lives, loss of biodiversity and loss of ethical values needed to build resilient societies. The large-scale degradation of the environment cannot be compensated for by token community projects. The so-called community development projects do not often address the problems. Shell, for instance, claims to have spent US$68 million in community development projects in 2007, but its share of the expenditure was US$20 million.[3] Note: because of a jaundiced joint venture arrangement, the Nigerian state picks up the bulk of the bills while Shell and its cohorts claim the credit. The entire expenditure, as is the case with the gas flare fines, are accounted for

1. Bassey, Nnimmo (2012). 'Beyond a Blackhole: Environmental Issues in Extractive Industries Transparency'. In Musa Abutudu, Dauda Garuba & Kolawole Banwo (Eds.), *Beyond Compliance; Deepening the Quality Assurance. Mechanisms of the EITI Process in Nigeria*. Abuja: Civil Society Legislative Advocacy Centre

2. Bond, Patrick (2006) *Looting Africa- the economics of exploitation*, University of Kwazulu Press, Pietermaritzburg/Zed Books, p67

3. Shell Sustainability Report 2007, p 24

as part of the production cost of crude. The communities are short-changed on all counts. This needs to be addressed.

Continued degradation in the form of oil spills and gas flaring and dumping of wastes render the Niger Delta extremely degraded. The area is also vulnerable to the impacts of climate change with a projected loss of 50 percent ability to produce cereals by the year 2020 and an 80 percent loss by 2050.[4] This is arguably worse than any armed conflict. TNCs must be reined in, held accountable for their historical debts to the region and made to take their heavy footprint off the land. For a start we demand a halt to any new oil field development in the region. The Niger Delta people will be better off as they can have clean creeks in which to fish and swim and also enjoy dark nights without explosive annoyances of toxic gas flares. The urgency of this resolution is more acute when we see that with peak oil, and the obvious shifts in energy sources that will happen, the future of crude oil is already history.

The Nigerian National Oil Spill Detection and Response Agency (NOSDRA)[5] announced in October 2009 that 2,122 oil spill incidents were recorded between 2006 and 2009. They estimated the amount of crude oil loss at 66,696 barrels. There were 252 incidents of spill in 2006; 597 incidents in 2007; 927 cases in 2008 while between January and June 2009, 346 cases were recorded.[6] The Nigerian government documented 6,817 spills between 1976 and 2,000, which according to analysts amount to one spill a day over 25 years. A 2007 report by Nigerian scientists and the World Conservation Union concludes that an estimated 1.5 million tons of oil has spilled in the Niger Delta ecosystem over the past 50 years, representing about 50 times the estimated volume spilled in the Exxon Valdez oil spill.'

It must be stressed that NEITI should not only count the income that comes from the sales of crude oil and gas. It is essential to count the costs of the negative impacts and inactions. Unless this is done, the figures arrived at in the balance books are largely off the mark.

It is for good reasons that the Niger Delta region of Nigeria has the reputation of being one of the most polluted places on earth. The observable evidence is enough to sustain that record and very little is in the public view with regard to detailed auditing of the devastation. The most comprehensive environmental audit of the region may be the Niger Delta Environmental

4. International Institute for Applied Systems Analysis (2008) *Food Security and Sustainable Agriculture – The Challenges of Climate Change in Sub-Saharan Africa*. At a side even of CSD16 at the UN, 8 May

5. NOSDRA was established by Act of the National Assembly on 18th October, 2006 an institutional framework to coordinate and implement the National Oil Spill Contingency Plan (NOSCP) for Nigeria in accordance with the International Convention on oil pollution preparedness, response and cooperation (OPRC) 1990, to which Nigeria is a signatory. The agency is mandated to ensure timely, effective and appropriate response to all oil spills as well as ensuring clean up and remediation of oil impacted sites.

6. Nigerian News Service reporting from Daily Independent (2009) 'Nigeria Records 2,122 Oil Spills In Four Years'. Tuesday, 06 Octoberhttp://www.nigeriannewsservice.com/nigeria-records-2122-oil-spills-in-four-years/ (accessed 7 June 2016)

Survey (NDES) commissioned by Shell Petroleum Development Company (SPDC or Shell) in the 1990s. That outcome of that survey has not been made public as the sponsor, Shell, chose to lock the reports up in their vaults.[7]

The crude path

The path of crude oil development has been soaked in human blood across the world and there is little to differentiate one oil company from the other when weighed on the scale of environmental responsiveness. Crude oil devastates both the environment and the unskilled local workers/contractors who are routinely exposed to the toxic substance. People exposed to crude, for example when there is a spill, are susceptible to suffering from digestive disorders as well as neurological problems.

Crude oil has remained a cheap energy source, even with the rising price in the market, principally because the environmental costs are overlooked in determining its value. The true price of crude oil has never been computed. The true costs would cover environmental, human, social, moral and security aspects of life in the oil communities. The environmental costs due to loss of intrinsic environmental services are incalculable in monetary terms. Oil spills; salinisation and siltation of water bodies; deforestation, gas flares and discharge of toxic wastes into the environment have made the oil communities death traps. When the environment is taken away from a people, you have effectively taken life away from them.[8]

Oil companies often open canals from the sea for the purpose of taking their equipment inland for extraction of crude oil. Canalisation brings salt water from the ocean and completely alters the ecosystems and overturns the means of survival of coastal communities. The people are forced to depend on the rain for potable water. However, because of insufficient rainwater harvesting systems, the people depend on wastewater from oil companies' facilities as evidenced at the Awoye community in Ondo State. The Awoye people know that the wastewater from the Chevron facility is toxic and have been told so by officials of the oil mogul. They, however, people insist that although they know that the water is poisoning them, they have no option but to drink it as they could not drink salt water. A glass of this toxic water looks like a glass of tea.[9]

Apart from the Biafra-Nigeria civil war[10], no other region in Nigeria has suffered the sort of sustained and inhuman aggression as the Niger Delta region. Nigeria recently celebrated 50 years of oil in the country. That could more aptly

7. Bassey, Nnimmo (2012) *To Cook A Continent – Destructive Extraction and the Climate Crises in Africa,* Oxford, Pambazuka Press

8. Bassey, Nnimmo (2008) 'Communities Subsidise Petroleum prices ... not Government,' Lagos, The Guardian, July 27

9. Bassey, Nnimmo. *Will they say our youths were not shot?* http://justiceinnigerianow.org/uncategorized/will-they-say-our-youths-were-not-shot (accessed 31 May 2016)

10. This war was fought 1967-1970

be termed 50 years of oil aggression in the Niger Delta. This assault has reached peaks in cases such as: the burning of Umuechem[11]; the massacre at Odi[12] and Odioma; the devastation of Ogoni and the extra-judicial murders of the peoples' heroes.

When asked to speak on the link between liberal globalisation and the acceleration of the destruction of the environment, Fidel Castro responded as follows, 'all efforts to preserve the environment are incompatible with the economic system imposed on the world, that ruthless neoliberal globalisation with the impositions and conditions by which the IMF sacrifices billions of people's health, education and social security...'[13] When asked whether the response to injustice in the world should be despair, he deduced that, 'The objective conditions, the sufferings of the immense majority of those people create the subjective conditions for the task of awareness-building.' The quote is long and worth the space, 'Everything is related: illiteracy, unemployment, poverty, hunger, illness and disease; a lack of drinking water, housing, electricity; desertification, climate change, the disappearance of forests, floods, hurricanes, droughts, erosion, biodiversity loss, plagues and other tragedies...'[14]

The conflicts are inevitably tied to the struggle for access to limited and receding resources, and the consequences of environmental degradation: deforestation, dislocation from territories, spills, gas flaring, ill-health, and poverty. In a context where the peoples of the region depend largely on environmental resources (potable water from the streams, creeks and rivers; fuel wood for energy; herbs for medicine; the land for agricultural produce and the creeks for fishing), environmental degradation has become a wholesale harbinger of death.

Crude footprints

The crude extraction chain can be said to commence at the exploratory stages when seismic lines are cut through forests and farmlands and creeks. Shell alone, in the two states of Bayelsa and Rivers, is said to own 56,000 kilometres of seismic lines. These lines mean losses of huge swaths of forests and farmlands. The corporations grab these lands courtesy of an obnoxious Land Use Act that the military set in place in 1978. The exploratory lines open the land for invaders and opportunists of all sorts who magnify the loss of biodiversity and environmental services in such areas. The seismic phase also involves the use of

11. In 1990 Umuechem community was sacked by mobile policemen and 500 houses were burnt down and 80 persons were killed simply because the people dared to organise a protest against the exploitation of their resources with no due benefits accruing to them.

12. The Odi massacre occurred in November 1999 under the presidency of Chief Obasanjo. 2843 citizens including children and the aged were killed. For details see ERA: *A Blanket of Silence*

13. Castro, Fidel with Ramonet, Ignacio (2007) *My Life*, London, Penguin Books, p397

14. Castro (2007) P399

explosives such as dynamite in test holes on the land and in the creeks. These further lead to losses of animal and aquatic lives.

The drilling stage involves the use of radioactive substances that are difficult to dispose of as they come out in wastewater as well as in drilling mud. These ought to be handled with utmost diligence, but it is believed that these are recklessly discharged into the Niger Delta environment, posing grave health hazards to the unsuspecting populations.

Oil spills are a frequent occurrence in the region. The corporations are good at claiming that these result mainly from sabotage and/or vandalism. This claim must be taken as an afterthought because even before the rise of armed confrontations in the region, over 300 incidents of spills were officially recorded each year while independent estimates were put at over 1,000 incidents. The spills and the manner of their cleanups are arguably violent attacks on the people. Due to a lack of knowledge, some communities have compounded this problem by insisting that they must receive compensation before any attempt is made at cleaning up such spills. They also often insist that the clean-up contracts should be given to local contractors who do not have the skills or the equipment to safely and adequately execute such tasks. This way, local people are exposed to danger due to haphazard handling of the toxic substances. And, because the people are involved with such cleanups they end up having to live with shoddy jobs that leave the environment worse than it ought to be. Cases have been recorded where forests have been set on fire in a bid to clean spills. Water bodies have equally been set ablaze in futile efforts to clean spills. Lives have been lost in spill related fires when community people use faulty machines to attempt to pump crude oils from spill sites.[15]

Some health impacts[16] related with oil spills and chemical substances used in the industry and to which the people of the Niger Delta are exposed on a continuous basis include:

- Blurred vision, eye-reddening
- Headaches
- Sore and bleeding
- Asthma, bronchitis and other breathing problems
- Increased risk of tuberculosis
- Ear infections
- Skin irritation and rashes
- Cancer (skin, lips, mouth, lungs)
- Menstrual problems, miscarriages, stillbirth, and birth defects
- Ulcers
- Heart attack
- Damage to liver, kidneys, etc.
- Lungs and throat infections

15. Environmental Rights Action field reports and testimonies

16. Oilwatch International Monitors Handbook. This may also be found in Conant, Jeff and Fadem, Pam (2008) *Community Guide to Environmental Health*, Hesperian Foundation, P.506

The challenge of the offshore

The many environmental conflicts that have occurred onshore over the decades appear to the encouraging the major oil companies to place their bets on offshore fields. However, the offshore has its peculiar challenges. Fields are being found in deeper waters and there are already technological challenges in handling accidents. A clear example can be seen in the April 2010 BP oil spill in the Gulf of Mexico.

That spill revealed some problems of transparency in the oil industry. It revealed that operators cut costs, care less about safety of the environment and evade the truth about their activities, among others. It also revealed that BP did not conduct genuine environment impact studies/analyses for the projects. They also did not have adequate oil spill response plans and mechanisms.

In terms of transparency, we saw the spill volumes increasing over time, as the company was forced to be more realistic with the figures. It was a shameful display of corporate duplicity and unwillingness to be open.

The figures released by BP changed as follows: 1,000 barrels per day as at 25 April, 5,000 barrels per day by April 28; 12,000— 25,000 barrels by May 27; 20,000 – 50,000 barrels per day by early June. The figures eventually hovered around 100,000 barrels a day.

The impacts on livelihoods in Nigeria are much more than what would be recorded in the Gulf of Mexico, for example. This is so because the communities depend almost entirely on the basic environment. When streams and ponds are polluted, potable water availability becomes a critical matter. Burnt forests mean denial of medicinal herbs and food. Fish and other aquatic populations die, and farmers lose their sources of subsistence as well as income because of soil destruction.

Some examples of spills in the Niger Delta[17]:

- The Escravos spill of 1978 in which 300,000 barrels of crude oil was spilled into the coastal waters
- Shell's 1978 spill caused by tank failure at Forcados Terminal in which 580,000 barrels were spewed
- Texaco's (Chevron) Funima-5 offshore blow out in 1980 that released 400,000 barrels of oil
- Mobil's spill at Idoho in 1998 with a reported release of 40,000 barrels of crude oil.
- The Shell in 2008 spill at Ikot Ada Udoh spill where a capped well failed and spilled an unreported amount of crude oil for months before it was stopped
- AGIP oil spills at Kalaba, Bayelsa State raged for over two months starting from February 2009 before it was stopped.
- Exxon oil spilled more than one million gallons (about 28,570 barrels) of crude oil from a ruptured pipeline in Akwa Ibom State

17. For detailed field reports of oil spills and other incidents see *Knee Deep in Crude* (Volumes 1 and 2) – a compilation of ERA's field reports. Environmental Rights Action, 2009

starting from 1 May 2010. The spill went on for seven days before it was stopped. As we speak, local people are still suffering from the impacts of the spill.
* Massive spills at Bodo, Ogoni, 2008/2009

We can also point at the oil spills at Gokana (2007), Aleibiri (1998) and Goi (2004) all attended with fire outbreaks that destroyed water bodies and forests. A list of spill sites would take up a whole book. By official records (NOSDRA) there are about 300 oil spills every year. In addition, we are informed that there are about 2400 oil spill sites that are yet to be attended to. This fact underlines the falsehood in industry propaganda when they claim that they do engage in clean-up of spills. If they do, how did 2400 spills accumulate?

Professor Richard Steiner of the University of Alaska made a comparative analysis of the practices of a particular company operating in Nigeria, Shell Petroleum Development Company, with international standards to prevent and control pipeline oil spills and observed that Throughout 50 years of oil production, this ecologically productive region has suffered extensive habitat degradation, forest clearing, toxic discharges, dredging and filling, and significant alteration by extensive road and pipeline construction from the petroleum industry. Of particular concern in the Niger Delta are the frequent and extensive oil spills that have occurred. Spills are under-reported, but independent estimates are that at least 115,000 barrels (15,000 tons) of oil are spilled into the Delta each year, making the Niger Delta one of the most oil-impacted ecosystems in the world.'[18]

Professor Steiner also observed that oil spills have a significant impact on the natural resources upon which many poor Niger Delta communities depend. Drinking water is polluted, fishing and farming are significantly impacted, and ecosystems are degraded. Oil spills significantly affect the health and food security of rural people living near oil facilities. Additionally, oil spills and associated impacts of oil and gas operations have seriously impacted the biodiversity and environmental integrity of the Niger Delta. Besides these direct environmental impacts and although the industry regularly claim that most of their oil spills are caused by sabotage, it is believed that the rate of spills per length of pipeline in the Niger Delta is much higher than is the case in developed countries such as the US. The conclusion reached by Steiner is that 'This, and other evidence, suggests that oil companies operating in the Niger Delta are not employing internationally recognised standards to prevent and control pipeline oil spills.'[19]

Shell announced a spill of 'less than 40,000 barrels' of crude oil escaping its pipe while loading an oil vessel at its Bonga Floating Production, Storage and Offloading (FPSO) facility on 20 December 2011. The company announced

18. Steiner, Richard, *Double Standards? – International Standards to Prevent and Control Pipeline Oil Spills Compared with Shell Practices in Nigeria.* (not yet published)
19. Steiner, as cited above

it was fighting the spills, deploying two aircrafts and five vessels in the effort. In a bid to assure the world that the spill was insignificant, the company also announced that over a couple of days fifty per cent of the spilled crude had naturally dissipated – meaning the crude evaporated or sank beneath the waves. The company's chemical assault on the slick was thus to fight a negligible errant sheen.

Satellite maps and photos published by SkyTruth helped to ensure that the world had a somewhat independent information on the spill apart from what Shell was announcing. Shell was largely in control of what people knew and said of the spill. Even officials of the Nigerian Oil Spill Detection and Response Agency (NOSDRA) did not come up with anything different from what the oil major claimed. The agency apparently swallowed the line Shell had cast in the waters that this spill was the largest when limited to a span of time not going back beyond a decade. This was a masterstroke by Shell's information managers, a coup of huge proportions. They almost succeeded in their game of enforcing a collective amnesia and deflecting focus to Exxon whose 1998 New Year's spill Shell appropriated as a benchmark.

The oil companies are masters at understating (and underreporting) the amounts of crude they release into the Niger Delta environment. We remind ourselves here that Shell spilled 570,000 barrels of crude oil at its Forcados terminal in 1979.

There was no independently verified record of what transpired at the Bonga FPSO. How much crude was actually spilled? What was the actual cause of the spill? What chemicals were used in tackling the spill? What were the impacts on the endemic Bonga fish species in the area and what does that mean to the food chain?

Before the fundamental JIV was carried out, Shell got busy flying government officials and journalists over the Bonga region to prove that they had contained the spill and that other reported spills were not from the Bonga FPSO. Reports from over-Bonga-spill-flights indicated that government officials refused to dance to beats from Shell's bongo and conga drums. Fishermen and community environmental monitors had informed Environmental Rights Action (ERA) and others that they spotted crude oil in Exxon's *Inanga* (name of endemic fish species) field off the coast of Akwa Ibom State, as well as in the Bisangbene River at Vanish Island in Odioma and St Nicholas areas of Bayelsa State. During the flight Shell reportedly sought to assure government officials that the thick crude they sighted on the waters was from a ship in the neighbourhood of the Bonga FPSO and not from Shell's activities. The officials reportedly retorted that they could not exonerate Shell and blame the accused vessel from the air. Thorough investigations are needed, they rightly said.

The period also saw spills recorded at Otumara in Escravos area of Delta State where locals said it had been ongoing for onward of two weeks without official records. More crude oil spills also hit River Ramos close to Escravos at about the same time.

The season of spills also had one from AGIP's facility at Okpotuwari in Southern Ijaw Local Government Area of Bayelsa State. AGIP officials visited the spill location on 27 December 2011 but told the community people that they would return the following week to stop the spill. For days the crude spewed unchecked into the Okpotuwari environment. The people had no respite from these and other environmental assaults but the officials of AGIP and other oil companies enjoyed their New Year vacation undisturbed.

Fires on the waters

Almost a month after the Shell spill, an explosion occurred on a gas well drilling rig at Chevron's Apoi North field on 16 January 2012.[20] That explosion killed two workers. Chevron, like Shell, offered initial information and updates on the incident, assuring that steps were being taken to contain and mitigate the disaster. Like Shell also this clearly was an half-hearted public relations gesture and tapered out even while the inferno raged.

Located a mere ten kilometres off the Bayelsa coast, and in fairly shallow waters, the glow and roars of hell competed with the numerous gas flares onshore. The impacts on the Koluama 1 and 2 communities, Ikebiri 1, 2 and 3 communities among others, cannot be dismissed or denied. The evidence floated and still floats on the waters. Dead fish including dolphins, and at least a whale, indicate that what has occurred was not a minor incident but a catastrophe.

As Chief Christian Munghanbofa-Akpele, chairman, Council of Chiefs, Koluama 1, told Environmental Rights Action monitors, this accident has raised serious concerns. He said 'We suffered a similar thing in 1980 when there was another major oil spill from Funiwa 5; just about 300 metres from the site that is on fire now; the Apoi North.' He asked 'Must we continue to have the negative impacts from Chevron's operation in our environment, without corresponding benefits?'

At Ikebiri 2, Mrs. Suoyo Matthew lamented:

It is very unfortunate that we are experiencing all these environmental devastation because of crude oil exploration. We have been suffering from series of oil spills from AGIP's facility. Now, just see what the recent incident from Chevron facility has brought us. Look at my body. I have blisters all over my skin. I feel very sad and uncomfortable these days. Just because of crude oil and gas exploitation around us we are facing a kind of ecological war; our livelihood and health are jeopardized. I really lack words to express myself. Government should help us address

---

20. Bassey, Nnimmo: 'Roaring flames on the bight of Biafra' at http://www.pambazuka.org/governance/roaring-flames-bight-biafraaccessed 31 May 2016

this issue and save us from poverty and life threatening activities of oil companies operating in our environment.[21]

One full month passed before Chevron started drilling a relief well in order to plug the damaged one. Soon after they commenced drilling, the raging inferno ceased. Community people reported that they could not see the usual 'orange glow' of the fire from the night of 2 March.

Chevron does not know, and cannot explain why and how the fire got extinguished. They hazard a guess that rocks may have fallen into the damaged well and plugged it. The dangers associated with this murky state of affairs is that since the fire has not been stopped in a controlled and efficient way, we cannot be sure there is no further leakage or that there would be no further eruption or explosion. While Chevron claims that there is no further leakage of gas, community monitors report that there are bubbles on the waters and noxious odour from the area.

There are many lessons to be learned from these incidents. First of all we learn that offshore oil and gas activities are accident-prone. A catalogue of these has been logged from around the world. We learn also that even in shallow waters the oil companies lack the readiness and capacity to handle these accidents expeditiously and effectively. The response mechanisms by the oil companies as well as government's regulatory agencies are dismal and the government agencies appear to be tied to the apron strings of the companies and lack independent capacity to act. This replicates around the world because of the revolving chair relationship between the two.

We learn also that there is an embarrassing lack of seriousness on the part of our public officials. The minister with oversight visited the Chevron accident scene and impacted communities one full month after the explosion.

The neglect of communities is legendary. When the Chevron fire impacted the communities, a protest of Koluama community women to Chevron's offices in Warri, Delta State, only resulted in the sending of token relief materials to some of the communities. Meanwhile fundamental issues of environmental remediation and restoration are not on the cards.

It is this sort of thinking that allows criminal exploitation and environmental despoliations to go unchallenged. The United Nations Environment Programme's report on the Ogoni environment remains unattended to, several months after its submission. That report is evidence of the lack of care of both the corporations and the government about the environment and the people. It echoes the cries of the Ogoni people and the peoples of the Niger Delta at large for a cleaning of the mess accumulated over the years of oil exploitation in the region. It is not too much to demand an investment of a bit of the capital accumulated from exploiting the communities and their environments for over more than half a century. When oil no longer

---

21. Nnimmo Bassey (11 March 2012) 'Fires on Our Waters'. http://nnimmo.blogspot.com.ng/2012/03/fires-on-our-waters.html

earns reasonable revenue, will anyone invest in cleaning the oil field communities? Sooner than we expect, the world will move away from dependence on crude oil. But the pollution will not go away unless they are cleaned.

Selective transparency

The Nigerian Extractive Industries Transparency Initiative (NEITI) is a vital tool in the hands of both the Nigerian government and the Nigerian people for ensuring transparency in the extractive sector. With the already published audits it has become clear that the oil and gas sector is particularly opaque. This situation is not peculiar to Nigeria as the sector fights hard to ensure that their books and deals are not subject to public scrutiny. Their usual plea is that revealing exactly how much they pay to certain governments may constitute a breach of business secrets that may lower their competitiveness in the field. While claiming a right to information blackout, the companies insist that they believe in the principles of the Extractive Industries Transparency Initiative. However, selective transparency is no transparency.

For a moment let us examine recent manoeuvrings by companies operating in the extractive sector as they squirm and resist basic transparency. Much of 2011 was spent by transnational corporations listed on the US Securities and Exchange Commission (SEC) in wrangling over clauses in a new financial reform aimed at ensuring that these entities discontinue operations on the well-worn path of double 'transparency' standards.

The US government proposed that mining, oil and gas companies who trade their shares on the American stock exchange should issue an annual report detailing the 'type and total amount' of payments they make to foreign governments. In the proposed amendment, Section 1504 adds Section 13(q) to the Securities Exchange Act of 1934 and this has kept the sector on their toes. This order is seen to be necessary if President Obama's Executive / EITI passes. In addition, the API does not 'believe it is necessary for the rules to specifically list other types of fees that would be subject to disclosure. We note that fees related to entry into, or retention of, licenses or concessions can be competitively sensitive information.'

The continuation of gas flaring in Nigeria has been aided by the jaundiced transparency in the sector. When the companies say they are doing everything possible to stop flaring they can be understood to be playing mere public relations. While they claim to wish to end the illegal act, these same companies are busy raising hurdles on the path of ending the unhealthy practice.

For example, Shell, ExxonMobil and Chevron are said to be deliberately frustrating government efforts to install real-time measurement equipment at 166 gas points to accurately meter the amount of gas being produced in the country. According to the Directorate of Petroleum Resources (DPR) only ten out of the 166 points have had the measuring equipment installed. This posture compounds the lack of transparency in the Nigerian oil and gas sector, where the true amount of crude oil extracted in the oilfields remains a mystery. Figures of gas being produced and how much is actually flared are mere guesses.

There should be no surprise that the World Bank states that gas flaring decreased in 2009 in Nigeria from 21.3 billion cubic metres to 15.2 billion cubic metres while Shell informs that their flares went up 33 per cent in 2010 over their 2009 figure. This clearly shows that whatever may have been the decrease in 2009 was not likely a result of the activities of the oil companies to curb the practice.

In terms of gas utilisation, more power plants would have come on stream by now, but investors complain that the flaring oil majors have generally refused to cooperate with them and deny them access to the gas that is currently being flared. In fact three years after thirteen companies prequalified by the Nigerian government to gain access to and harness gas from 180 identified onshore and offshore flare sites, the oil companies have not granted these companies the needed access.

Polluting bush refineries of the Niger Delta

Dire situations often inspire invention. Biafra was blockaded and starved of access to resources ranging from domestic goods to industrial products during the civil war. Necessity thrust upon Biafra the need to innovate and to create. It was in this mode that the nascent nation built and ran crude oil refineries and also produced missiles or bombs then known as 'ogbunigwe' or 'Ojukwu buckets'. These efforts were driven by the inescapable urge for survival.

In the past few years there has been an emergence of what many term 'bush refineries'[22] in the oil fields of the Niger Delta. These are spots in the swamps and creeks where local people, mostly youths, produce petroleum products using crude oil obtained from either already leaking pipelines or from spots broken into by crude oil thieves.

These refineries pose serious health hazards to their operators as they have no clue about the toxic nature of the products and do not have any sort of protective clothes, boots or gloves. These young folks bear the extreme heat from the flames of the belching dragons in order to produce litres of semi-refined products that pose additional threats to the end users. Many deaths related to kerosene explosions have been recorded and these may have resulted from the use of the uncontrolled products from these contraptions.

The dire poverty in the oil region is often cited as justification for the existence of these bush refineries. Regrettably, the response from the government, as well as from the political parties seeking control of the federal government after elections next month, is nothing beyond the provision of

---

22. *The Bush Refineries of the Niger Delta* published in the Oil Politics column of the author in 234NEXT. http://nigeriang.com/money/oil-politics-the-bush-refineries-of-the-niger-delta/8332/ (accessed 31 May 2016)

physical infrastructures in the region. While these are essential, the most urgent need of the region, and indeed the entire nation, is the detoxification of our environment. As we have often argued, the average Nigerian will take care of her basic needs if the physical environment supports her livelihood–generation efforts. This means that the urgent first step is an urgent audit of the environmental situation of the region as could possibly be exemplified by the assessment of Ogoni environment by the United Nations Environmental Programme (UNEP). More on this below.

A factor that could be perpetuating the bush refineries is the dislocation of the social infrastructure of the region. This includes the loss of communality, the rise of individualism and the deep corruption that has been entrenched by key players in the oil industry sector. These systemic ruptures must be structurally addressed.

We cannot ignore the efforts of security agencies in combating the menace of the illegal refineries. But merely combat posturing only gives the trigger-happy security men cover for extortion and further human rights abuses of an already traumatised people. However, it must be acknowledged that the continued operation of these bush refineries is a disservice to the local people and a huge shame to the government.

Going by figures from the Joint Military Taskforce operating in the Niger Delta, hundreds of these bush refineries have been destroyed. By mid-December 2009, the JTF reported that there were over 1,000 'illegal refineries' in the Niger Delta and that within two months to that time they had destroyed 600 of the refineries in different parts of the region. Sarkin Bello, the General who commanded the JTF at that time, made an important point that just as other ills had started in one part of the nation and spread to other parts, there was a chance that such refineries may pop up in other areas of the country – especially those through which oil pipelines passed.

Months later, Sarkin Belo bemoaned the resurgence of the bush refineries, as was widely reported in the mass media. It was not exactly surprising when a fortnight ago, the JTF announced that they had detected 500 bush refineries in the Mbiama area on the border between Rivers and Bayelsa States. It was not surprising because the refineries have been operating more or less brazenly with law enforcement agents sometimes accused of exacting tolls or illegal taxes from the operators. So they probably destroyed 600 in 2009 and the ghosts of the levelled plants resurrected soon as the security agents left the scene. These bush fires are huge tourist attractions for foreign journalists and you do not need a space rocket to gain access to their locations.

We have heard some politicians claim that the bush refineries cannot be eliminated because the youths cannot find alternative avenues of employment. Quite specious, that form of reasoning. It is illustrative of the ineptitude of persons in power who ought to provide employment and keep people away from practices that are harmful to them, the environment and the economy.

There are untold dangers related to operating these bush refineries. The poor youths who work these refineries, covered in crude, standing in the searing heat and continually inhaling toxic elements can hardly be in a position to enjoy the fruit of their labour. These refineries may put some kobo in their pockets, but they are essentially condemned to poor health and truncated lives.

A point that we must underscore is the fact that despite the large number of these bush refineries and the fact that they refine products that are illegally obtained, their operations do not lead to a reduction of the crude oil output of Nigeria. Why is this? It is simple to see.

Large-scale illegal bunkering with international dimensions has gone on unchecked for decades and many well-connected persons may be benefiting from it. The large-scale crude oil theft in Nigeria has gone on alongside and continual meeting of the production quota of the nation. The bush refiners may have been inspired by the fact that between the oil wells and the export terminals is a bottomless pit in which thievery is highly rewarded. Efforts at halting the petty stealing for bush refining will not be successful if the cancer of mass oil theft by the high and mighty is not tackled.

Rather than fight the menace in a structural way, oil companies seeking ways to heap the blame on victims have turned these refineries into some spots for pollution journalism tourism by flying foreign and local press corps over them.

The increasing pollution from these bush refineries must be tackled. So far the response of government has been to bomb and ransack the locations. The results have not been effective because not long after their demolitions, they come back to life. A more effective means of tackling the menace may be for them to be recognised as micro refineries and for regulatory measures to be put in place to ensure standards. That way, the crude would have to be bought and the refining standards would be enforced. This would help stop the spate of explosions that occur when ill-refined products are used in cook-stoves as well as halt the wastes and dumping of products into the environment simply because they were stolen in the first place.

Agonizing Ogoni Environment

When the Ogoni people demanded a halt to the unwholesome acts of the Shell Petroleum Development Company (SPDC or Shell) and the Nigerian National Petroleum Corporation (NNPC) the government called them names and unleashed security agents to maim, rape and murder and hound many into exile.

The report on the pollution of Ogoni land prepared by the United Nations Environment Programme (UNEP) was released on 4 August 2011. It marks the first official confirmation that there is a major tragedy on our hands. UNEP's report unequivocally shows that the Movement for the Survival of the Ogoni People (MOSOP) under the prescient leadership of Ken Saro-Wiwa was not crying wolf when it maintained that grave injustice was being inflicted on Ogoniland.

The report largely says what has been known and said before. But this is official and very valuable. When Shell doled out the funds for the study, they claimed they did so on the basis of the polluter-pays principle. True. Shell polluted Ogoniland, just as they and other companies have done and continue to do all over the Niger Delta.

Claims by Shell that a majority of the oil spills in Ogoni are caused by interference by local people flies in the face of the observations in the UNEP report. The report says the bush refineries, for example, became prominent from 2007. Obviously, one of the conclusions should have been that with livelihoods utterly destroyed, some of the people had to find a means of survival and chose this unfortunate and illegal trade. With UNEP's obvious care not to antagonise Shell in the report, this path was not pursued.

In a critique of the UNEP report, Prof Richard Steiner of Oasis Earth organisation, Alaska states:

> The UNEP report devotes several pages (161-166) specifically to artisanal refining at the Bodo West oilfield, and correctly reports an unfortunate increase in such between 2007 and 2011. However, in this analysis of oil pollution in this region, UNEP entirely ignores the other much larger source of oil spilled into this same region in that same time period— the twin ruptures of the Trans Niger Pipeline (TNP) caused by SPDC negligence in 2008 and 2009.Together these spills contributed between 250,000 – 350,000 barrels of oil into this system, orders of magnitude more than illegal refining. Much of the oil at Bodo West area likely derived from the TNP Bodo spills.[23]

How does this compare to the volume of spills from artisanal refineries?

Steiner also wonders why the UNEP study report says that no single clear and continuous source of spilled oil was observed or reported during UNEP's site visits,' whereas the massive spills at Bodo occurred at the time of the study and the combined spill volume may well exceed that of the Exxon Valdez that occurred in Alaska in 1989.

A significant problem that may scuttle efforts at acceptable cleanup of Ogoni land is the lack of capacity or unwillingness of Nigerian regulatory agencies to enforce laws and to act independently. Their independence is of course affected by the fact that Shell has infiltrated the petroleum ministry in a deep and total way.[24] If government is serious about regulating the sector it will need to ensure that those called to make this happen are not connected to Shell's umbilical cord.

How for instance could government officials certify that oil spills have been cleared up and impacted areas remediated whereas the contrary is the case? According to UNEP there are 10 'remediation completed' sites showing

23. Quoted in Bassey, Nnimmo (2012): To Cook a Continent: Destructive Extraction and the Climate Crisis in Africa. Oxford, Pambazuka Press
24. See WikiLeaks cables on the subject

ongoing pollution in Ogoni. Shell's spill management was also called to question as they use incompetent contractors for jobs that require knowledge, skills and equipment.

The confirmation that Shell has poor diligence in its oil spill responses and that our regulatory agencies endorse the pattern raises serious issues about the situation in other parts of the Niger Delta where this impunity continues unabated.

Other matters arising from the UNEP report that call for immediate follow-up include the inconclusive study on public health issues even though a gamut of medical records were surveyed. Same about vegetation and also rainwater that the people turn to in the face of living besides polluted rivers, creeks and waterways.

We now have official confirmation that the Ogoni people are drinking water polluted with benzene 900 times above World Health Organisation's standards. We also now know that the ground is polluted up to a depth of five metres at some places. We know that there cancer causing elements in the water and in the air. We also know that there are toxic wastes dumped in unlined pits in Ogoni land. These issues are replicated all over the Niger Delta. But they are heightened in those areas because you must factor in the highly toxic gas flares.

Ogoni land (read Niger Delta) ranks as one of the most polluted places on earth. What is urgently needed is for the Federal government to declare an environmental state of emergency there. Ecological problems do not observe community or political boundaries. How the government handles this case will tell a lot about who we are as a people.

Gas flaring and other wastes

Extractive businesses impact the environment at every stage of their operations. That is, impacts begin to occur and to manifest right from the exploratory stages except where such is being carried out through remote sensing including from satellites in space. However, wherever such explorations are terrestrial, their impacts are undeniable.

In an environment where regulations are weak and enforcements even weaker, it does not require much analysis to see that our environment is degraded to a large degree by the myriad toxic wastes and chemicals dumped into the land and waterways on a regular basis. One estimate has it that at least 600,000 barrels of water are dumped into the Niger Delta environment daily through oil production activities.

Health problems from millions of litres of produced water, drilling mud, etc., include those emanating from the pollution of the food chain and waters. For example, ingestion of polluted food and water from the swamps cause vomiting, dizziness, stomach ache and cough. *There is a case where within two months of dumping toxic wastes in a community, 93 people died from what* *the people termed mysterious illnesses.*There are also heightened and poisonous concentrations of lead, zinc and mercury, etc., in these pollutants.[25].

Another key by-product of oil production is *associated gas* or gas that comes out of the oil wells as crude oil is being extracted. This gas can be harnessed for utilisation for power generation or other uses, it can be re-injected into the wells or it can be set on fire when brought to the surface of the earth. The first two options require some investment while the last option requires virtually no costs on the part of the oil companies. The burning of this gas is generally termed gas flaring. The truth is that this is cheap to the companies but costly to the environment and to the communities living in them. In other words, when oil companies flare gas, they are simply avoiding responsibility and externalising the costs to the environment and the local people. Without taking these externalised costs into account, the true cost of oil cannot be known. By externalising these costs the companies and partner governments pile up ecological debt and poison the people and the environment.

Environmental and financial responsibility is clearly linked in the case of gas flares. Nigeria ranks second to Russia as a global gas flarer in terms of volume and by share of global gas flared Nigeria's contribution is 16 percent[26]. The estimated amount of gas Nigeria flares in 2,000 was put at 17.2 billion cubic metres[27] while other estimates place this at 23 billion cubic metres. It is important to note that gas flaring has been illegal in Nigeria since 1984 with the introduction of the Gas Re-injection Act. From the time that law came into effect, fines were introduced for gas flared and even then continuous gas flaring was to be permitted on a case-by-case basis by the responsible federal minister. Following a suit brought by Jonah Gbemre against Shell Development Petroleum Company for flaring gas in Iwerekhan community, a High Court sitting in Benin City declared on 14 November 2005 that gas flaring was both illegal and an affront to the human rights of the people and should be halted forthwith.That court order has neither been vacated nor obeyed as we write.

Over the years deadlines for the stoppage of gas flaring have been set through executive pronouncements and were never enforced. By the end of 2009 three deadlines were on the table, illustrating the impunity and irresponsibility of the sector in the country. While the Senate set a deadline of December 2010, the Executive arm set December 2011 and the oil companies snidely stated that only a 2013 deadline could be tenable. Later the same month both industry and government operators began to signal that having a deadline was of no use as they intend to create gas utilisation projects that they will seek to register for carbon credits through the Clean Development Mechanism

25. For more on this see HOMEF's Community Guide to Environmental Monitoring.https://nnimmo.files.wordpress.com/2015/12/community-monitoring-guide-homef.pdf (accessed 31 May 2016)

26. http://www.gao.gov/new.items/d04809.pdf (accessed 31 May 2016)

27. Ibid

(CDM) of the United Nations Framework Convention on Climate Change (UNFCCC).

In a rather unfortunate turn of events, the then director of the Directorate of Petroleum Resources (DPR) stated in a newspaper interview that the cessation of gas flaring would ruin the Nigerian economy.

Gas flares release a bunch of toxic chemicals into the atmosphere besides vast amounts of green house gases. . In addition to these, due to the inefficient combustion of the flames, a high amount of soothe is continually released into the atmosphere adding to the pollution problems. Gas flares are known to cause a variety of cancers, skin diseases, blood disorders, bronchitis, asthma and others.

In economic terms, Nigeria wastes US$2.5 billion in gas flares annually. We note that Nigeria has a huge reserve of natural gas and that efforts have been made to utilise some of those. Oil sector operators have equally tried to obfuscate matters by claiming to be using associated gas when indeed they are utilising non-associated gas from purely gas fields. A case in point is Chevron's futile effort to pass off the majority of the gas in their West African Gas Pipeline project. It has been estimated that not up to 20 percent of the gas conveyed by the pipeline is associated gas.

Before the shifting of deadlines game became more intense, oil corporations appeared to be serious about ending the obnoxious act of gas flaring. At a point Exxon had a target date of 2004, Chevron 2006 and Shell pointed at 2007. It must be noted that the World Bank prepared a frequently cited study of 2004 that documented the scope of the problem of gas flaring in Nigeria. By the time the carbon spin doctors began to speculate on carbon trading, the World Bank, which positions itself to be the major climate finance bank, is now keen on benefiting from the politics of gas flares. The bank is working with Nigeria to ensure that other countries can offset their emissions by investing in ending gas flaring in Nigeria. This is a clear case of why market mechanisms provide false solutions to climate change, potentially compound the problems on the ground and allow polluters to keep polluting. It is estimated that the World Bank will grab 13 percent of carbon offsets transactions. In the Nigerian case, the bank would make US$10/ton of gas that would have otherwise been flared.

The above scenario indicates that the environment is a blind spot when oil industry costs are computed. Key players and banks working in the sector focus on the profit and the environmental costs and social, economic and political costs are not countenanced.

Reserving our comments

Nigeria's crude oil reserve is put at 36 billion barrels while there is an estimated reserve of 100 -170 trillion cubic feet of natural gas. In 2004 Shell overstated their reserves by 4.47 billion barrels and this was faulted by the US exchange commission and was fined. In 2008 Shell reviewed its reserves downwards by

200 million barrels. On what basis does Nigeria check the transparency in this sector when basic information cannot be trusted?

The unreliability of crude oil reserve figures is further compounded by the rapacious stealing of crude oil going on daily with obvious complicity of local and international operators. The crude is stolen with sophisticated equipment, not mere buckets and shovels. The stolen crude is sold internationally and not to local refineries that chronically work at sub-optimal levels.

Although figures for stolen crude cannot be ascertained, we have had public figures speculate that as much as the amount that is officially exported daily is also shipped illegally. The governor of Delta State of Nigeria, Governor Emmanuel Uduaghan, has been quoted as saying that oil companies were involved in the illegal activities. He also said that the international community was complicit in the thefts since there was a ready market for the stolen crude.[28] For Mr Dimeji Bankole, speaker of the Nigerian House of Representatives, about half of Nigeria's crude oil production is stolen. He summed up that if that estimate is correct, it means that Nigerian crude may run out sooner than expected.[29] Other analysts believe that Nigeria is actually losing as much crude oil as it is selling officially through the connivance of security agencies that are meant to halt such practices. Through this the country is estimated to lose US$1.6 billion to oil thieves annually. Reports allege that some top naval officers, serving and retired, have private pipelines that run from the Port Harcourt area to Eket and that these pipes serve as conduits through which they siphon crude oil, load onto vessels and ship to refineries in other shores including South Africa.[30]

It is obvious that the fact that Nigeria depends mostly on crude oil and gas for revenue, and because governments over the years have grown away from engaging in productive activities, this sector has been allowed a free hand to inflict a reign of terror on the environment. NEITI needs to address this gap and expand the definition of transparency in the extractive industry to include environmental responsibility and not stay restricted to fiscal concerns. This should be extended to areas where solid minerals have been extracted and are being extracted. The Jos tin mines are still crying for remediation. The toxic artisanal mines of Zamfara State are still crying for comprehensive attention. The coal mines of Enugu have shockingly been turned into refuse dumps at places.

NEITI would render a great service if it helps Nigeria to see whether her accounting books are in the black or red. The conclusion of this chapter is that there is a huge transparency black hole in the sector. Beyond just having this

28. Owuamanam, Jude (2009) *Bunkering: Uduaghan Blames Oil Firms, Multinationals.* Lagos: The Punch. November 8, Page 5.

29. Ojo, Eric (2009). *Bankole laments illegal oil bunkering in Nigeri Delta – challenges security agencies on leakages in public funds.* Lagos, Businessday, November 8, page 8.

30. Ehirim, Chuks (2009) *How Naval Officers aid Bunkering,* Lagos: National Daily, November 16-22

reality is the urgent need to audit the environment and urgently embark on a Niger Delta and indeed a national environmental restoration.

23

Drilling in the dark

This blog piece was first published on 234NEXT on May 5, 2011[1]

The Nigerian oil sector must be one of the sectors that tolerate blatant disregard for transparency in the land. Being a mono-product economy and depending so much on foreign expertise, technology and dictates opens the sector to peculiar challenges.

A reading of the 2005 Nigerian Extractive Industries Transparency Initiative's (NEITI) audit report[2] reveals three interesting things. One of them is that the Niger Delta Development Commission (NNDC) claimed to have received more money than it was given. There must be more miracles lurking in the accounting books of the NNDC. Remember that in their 2010 budget, they had a chicken-change sum of N90m for staff marriages and bereavements! The commission defended the outrageous budgetary allocation on the grounds that it was dictated by emotional intelligence. Peculiar intelligence, one would say.

The second interesting matter that emerged from the NEITI audit was that the Nigerian National Petroleum Corporation still relies largely on paper-based accounting systems. This could be a possible reason why we keep receiving conflicting signals as to whether the corporation is solvent or insolvent. Besides cracking our brains over the incoherence that reigns in the chambers of the executive council, we should perhaps pardon ministers and big shots who have been shooting out those divergent messages. If you have to drill through all those piles of paper, with figures backed with endless zeroes, at the end of the day, you could end up at any end of the pipe. And, who knows, some rats may help themselves to some of those delicious crude covered accounting sheets. Some calculators were said to have become overheated during election figures collation simply because they were not given enough time to cool down before new figures were hammered in.

The third thing we will consider should receive the gold medal for crass impunity. The NEITI auditors reveal that Nigeria does not know exactly how

1. http://www.legaloil.com/NewsItem.asp?DocumentIDX=1304689984&Category=news (accessed 31 May 2016)
2. See at http://neiti.org.ng/index.php?q=pages/2005-audit-report (accessed 31 May 2016)

much crude oil is being drilled from the many wells of the Niger Delta on a daily basis. The operators, the oil companies who often claim to be baking the national pie, would simply not provide such data to the auditors. But they do provide some sort of figures, don't they? Of course. They give figures of how much crude reaches the export terminals and other distribution points.

The question is: what happens between the pump heads and the terminal points? The massive leakage that occurs between those points is what the oil companies do not want us to know. That gaping hole is what the Nigerian government must plug. That sore gash is what the Nigerian people must demand an account of.

Reports are replete in the news media of petty oil thieves in the creeks of the Niger Delta who break pipes, siphon crude oil into drums and tanks and then refine them in rickety contraptions often referred to as illegal or bush refineries. While no one can deny the existence of these pilferers, the truth must be told about where the bulk of Nigerian crude goes and into whose throats and pockets.

Why would the oil companies refuse to give figures of extracted oil measured at the well heads? Why is the Directorate of Petroleum Resources (DPR) unable to independently measure and provide such figures? Who are those raising brick walls against transparency? Why are we prostrate before the altars of these oil moguls? We have heard of offers being made to the DPR to acquire equipment as well as training for independent metering of production in the oil fields. What or who stopped the acceptance of that much-needed capacity boost?

The clear suspicion in all these is that the oil companies are complicit. There must be something to gain by hiding the figures. Pronouncements from public figures such as the outgoing Speaker of the House of Representatives and the governor of Delta State, among others, add up to mean that probably as much oil as is being officially exported daily is also being stolen.

Remember that a ship caught with stolen crude sprinted out of naval detention a couple of years ago. We perceive a matrix of high-powered players in the oil theft industry. This is far beyond pointing fingers at petty thieves who steal crude oil in buckets only to ferry them in crude barges to ships lurking off the coast. An international syndicate must be at play, with local fat cats keeping the machines well oiled, literally.

The NEITI Act empowers the body to prosecute any company or government official who refuses to give needed information, or who falsifies the information that may be needed in the furtherance of the pursuit of transparency in the sector. Not knowing exactly how much oil is being extracted daily raises a number of concerns. For one, we cannot reasonably be sure of how much Nigeria's oil reserves are if the amount being extracted is not known. Secondly, we cannot reasonably estimate how much crude oil is being stolen or lost into the environment.

Why no oil company or government official has been prosecuted for refusing to tell Nigeria how much crude oil is being drilled on a daily basis, is a question that needs an answer. We simply cannot keep on drilling in the dark.

24

So Shell is everywhere

This piece, published in 2009, responded to how Shell influenced public policy in Nigeria[1]

It was bound to bubble to the surface one day, that the multinational oil companies operating in Nigeria had a certain foothold on the Nigerian government that is more than having a toe in the door. The WikiLeaks reports showed a brash Shell official boasting of how they have infiltrated every facet of the Nigerian government.

This should, however, not surprise anyone. Did they not draw up Nigeria's Vision 20/20 under the Abacha/Shonekan regime? When the government broached the idea of a new oil sector bill, didn't Shell's Ann Pickard, the then vice-president for sub-Saharan Africa, warn that the oil company would not accept any law that is against the interest of the company? And that was stated at a meeting in Abuja, not in the creeks of the Niger Delta; yet no government official made even a whimper in protest of such an affront on a sovereign state by a company.

Miss Pickard was then quoted by the Financial Times[2] (24 February 2009) as saying 'We do see that the legislation, the bill, will have a profound impact on the way the industry functions and how the companies move forward...Getting it right [is] absolutely essential. Getting it wrong will not be acceptable for Nigeria or the [oil companies].'

WikiLeaks tells the world that Shell had intelligence to share on militant activities as well as on business competition in the Niger Delta. We are also told that Shell knows how leaky the Nigerian government is. What a sorry picture the then minister for petroleum resources, Odein Ajumogobia, must have cut when he denied a letter from the government inviting China to bid for oil concessions.

Sneaky Shell already had a copy of the letter. And they also knew that similar letters had been sent to Russia, according to WikiLeaks. The meeting

1. http://www.shelltosea.com/content/oil-politics-so-shell-everywhere (accessed 31 May 2016)
2. Green, Matthew (25 February 2009) 'Shell warns Nigeria over oil and gas reforms', FThttp://www.ft.com/cms/s/0/6ae2e146-0276-11de-b58b-000077b07658.html#axzz43jF135fpaccessed 31 May 2016

with the Russians was even recorded, transcribed and sent to Shell. Interesting, but not surprising. These Shell spies must be so trusted and well paid by the company otherwise one would have asked if the transcripts were accompanied by sworn affidavits.

Shell's Pickard is quoted as saying to the US ambassador that 'the GON [government of Nigeria] had forgotten that Shell had seconded people to all the relevant ministries and that Shell consequently had access to everything that was being done in those ministries.'[3]

It can be suggested that today, with a former Shell director sitting as the minister of petroleum, Shell may not need small fries to snoop and scan pages from that ministry's bulging filing cabinets. They may not have to rely on low level officials with tape recorders concealed in pens, tie clips, belt buckles, eyeglasses, or cufflinks to record meetings and send transcripts to them. Now they may have copies of whatever document they want forwarded directly as a matter of routine. Hopefully, that would not be the case.

The game of infiltration of public office by oil companies is not limited to Nigeria. It was revealed in the BP Deepwater Horizon fiasco that regulatory agencies in the USA were very chummy with the oil mogul's official and that this contributed to the lax oversight. How else would BP have claimed several times over, in the oil spill response plan as well as their environmental impact assessment that there were virtually no risks associated with such an operation?

In BP's Deepwater Horizon exploration plan[4], the company specifically stated in Section 10 that A description of the measures that would be taken to avoid, minimize, and mitigate impacts to the marine and coastal environments and habitats, biota, and threatened and endangered species is not required.' The company went ahead to say in Section 14 that 'No adverse impacts to endangered or threatened marine mammals are anticipated' and that also 'No adverse impacts to endangered or threatened sea turtles are anticipated'; and 'No adverse impacts to marine or pelagic birds are anticipated'.

Impunity is the word in this sector. Disrespect of the sovereignty of nations is the norm. How do you regulate companies who play the game by any rule they chose to set to ensure their dominance and profiting? How do you regulate an industry that engages, as suggested by these leaks, in espionage possibly under the guise of business research?

The response of the NNPC's spokesperson, Levi Ajuonoma, as published in The Guardian (London) is pitiful. He is quoted as saying that 'Shell does not control the government of Nigeria and has never controlled the government of Nigeria. This cable is the mere interpretation of one individual. It is absolutely

3. WikiLeaks: Shell 'knows everything' about Nigerian government. (09 December 2010), Telegraph, UK. http://www.telegraph.co.uk/news/worldnews/wikileaks/8190406/WikiLeaks-Shell-knows-everything-about-the-Nigerian-government.htmlaccessed 31 May 2016

4. See Deepwater Horizon: Disaster in the Gulf - AN OIL RIG CALLED DEEPWATER HORIZON. https://www.awesomestories.com/asset/view/AN-OIL-RIG-CALLED-DEEPWATER-HORIZON-Deepwater-Horizon-Disaster-in-the-Gulf//1 (accessed 31 May 2016)

untrue, an absolute falsehood and utterly misleading. It is an attempt to demean the government and we will not stand for that. I don't think anybody will lose sleep over it.'[5]

It is true we have lost so much to the activities of Shell and other oil companies in Nigeria, including the NNPC. We have lost lives, our environment and our dignity. We can say that we are tired of losing things to this sector. However, to not lose sleep over this revelation of the dealing between oil companies and embassies and our government circles is to ask us to shut our eyes to a dangerous travesty.

5. Smith, David(8 December 2010), WikiLeaks cables: Shell's grip on Nigerian state revealed. The Guardian (UK) http://www.theguardian.com/business/2010/dec/08/wikileaks-cables-shell-nigeria-spyingaccessed 31 May 2016

25

Shell's fracking moves in the Karoo

This was first published on April 21, 2011 on 234NEXT[1]

There are some words that those who develop dictionary software appear somewhat slow to catch up on. One of such words is 'fracking'. While the word is still kept on the fringes of everyday discourse, the process it describes is already pitting citizens against corporate power in North America, Europe, and in Africa.

As the sound of the name suggests, fracking has to do with fracturing. The New American Oxford dictionary defines fracture as 'the cracking or breaking of a hard object or material ... a crack or break in a hard object or material, typically a bone or a body of rock...the physical appearance of a freshly broken rock or mineral, esp. as regards the shape of the surface.'

Fracking has already raised serious problems in the United States and is being questioned and resisted elsewhere. The nearest flash point is the resistance to Shell in their efforts to engage in fracking in the Karoo, South Africa. The community resistance in South Africa is especially interesting in the sense that Shell has been confronted there by their Nemesis: Ogoni activists displaced by their activities in Nigeria.

In the case of the plan by Shell for fracking in South Africa, they plan to bore holes 5 kilometres down into the belly of the earth in order to extract gas trapped in a layer of shale stones. This is another signal that the age of cheap oil is over.

Fossil fuels are being sought for in increasingly less accessible locations such as deep-water locations and in locations previously considered off limits to extractive activities. As someone said, some of the processes can be likened to a 'societal scraping of the barrel.'

This process is not exactly new, as it has been going on in the USA for decades, according to some records. The causes of current anxieties are primarily two-fold. Companies involved in this business have not released the names and quantities of all the chemicals they use in the fracking processes.

Secondly, the process uses huge amounts of water, a serious concern in a season of water scarcity. After pumping in huge volumes of water, about half of

1. No longer accessible

this water is pumped out and the bubbles or gas are removed. The wastewater with all its highly toxic dregs is then disposed of. The question is whether this is handled in a manner that assures of safety.

According to the experts, Shell's proposed exploration will apparently entail drilling 8 boreholes in each precinct (i.e. 24 boreholes in total) of up to 5 kilometre depth over a three-year period, extendable to nine years.

'It appears that each well will need between 0.3 million and 6 million litres of water (i.e. a scenario of between 7.2 million and 144 million litres of water required). Shell has been extremely vague as to its anticipated source of water, with no concrete indication being given in the draft EMP or in the public consultation meetings as to where the multinational intends to source the requisite water from.'[2]

While some people argue that there are yet. to be analyses showing actual water contaminations related to chemicals used in fracking, there are several confirming water contamination due to fracking processes.

For one, some of the chemicals used in the process are known as carcinogens. The US Environmental Protection Agency is examining the potential impacts on drinking water of the various stages in the hydraulic fracturing process. Such stages include when drillers mix water with chemicals and sand and inject the fluid into wells in order to release oil or natural gas.

Some 46 House of Representative Democrats sent a letter to the Secretary of Interior in which they stated, 'Communities across America have seen their water contaminated by the chemicals used in the hydraulic fracturing process.'

Other concerns over fracking plans have been raised in Canada and France. A report from the Tyndall Centre in the United Kingdom, and an enquiry by the House of Commons, has trailed the fracking business in that country.

The Tyndall Report found a paucity of information on which to base serious analysis 'of how shale gas could impact on GHG emissions and what environmental and health impacts its extraction may have; that there is a clear risk of contamination of groundwater from shale gas extraction.'

Fracking folks have enjoyed exclusion from regulation in the USA for years and are very reluctant to accept accountability today. With Barack Obama's intent to accelerate the weaning of his country from heavy reliance on crude oil imports, the shift to fracking seems good to some investors, irrespective of its highly toxic and water-guzzling nature.

The exportation of that anti-regulation operational latitude to other lands is meeting serious resistance. The people of Karoo are basing their resistance, among other things, on the indelible footprints that Shell's operations etched into the hearts, veins, and blood of the Ogoni.

The linkage between the Ogoni and the Karoo deserves an applause as ordinary people rise up to ask, 'what the frack is going on' and link hands

---

2. Hevemann Inc. A critical Review of the Application for A Karoo Gas Exploration Right by Shell Exploration Company BV. http://www.imel.uct.ac.za/sites/default/files/image_tool/images/315/Research/Research_by_Staff/FracturingLegalReportApril2012.pdf p.3 (accessed 31 May 2016)

across political boundaries to globalise the struggle and hope for the security of humankind in a globalised world.

26

The coming belt of fire

This was first published in 234NEXT on August 05, 2010[1]

On 20 April the world woke up to what oil spills mean and could mean. Many reporters and the news media suddenly realised that there were heavy spills in the Niger Delta, besides the gas flares that toast the skies daily.

However, even as the media lenses are focused on some of the atrocious evidence of environmental impunity in our backyard, the angling for new oil blocks is assuming a stronger beat in the corridors of Aso Rock, as well as in the board rooms of oil companies and related speculators.

The government needs more revenue; the oil companies need more profits— it is a crude wedlock of convenience. Meanwhile the people are crying for mere space for survival. Who listens to them?

Furthermore, as crude oil reserves deplete, oil companies are moving into more fragile environments: off shore and even eco-reserves. There are also more concerted moves into dirtier forms of crude— such as bitumen development.

Bitumen mining produces three to five times more greenhouse gases than conventional crude oil extraction. With the plans by government to exploit bitumen from Edo state to Lagos state, we can expect a belt of fire in this region that will make the Niger Delta conflict a weak prelude.

Bitumen is extracted largely by two methods: open cast mining or drilling somewhat like crude oil is extracted. The open cast mining system means excavation of the soil to reach the mineral necessitating the uprooting of everything in its path. This means that whereas communities have been polluted in the Niger Delta, in some of the areas where bitumen will be mined, communities will simply have to be relocated or just dislocated. Where bitumen is to be extracted by drilling, steam has first to be pumped into the wells to melt the mineral and thus make it possible to pump to the surface through pipes. All these add to indicate that bitumen belt will indeed be a belt of fire.

Even though a monster cap has been fitted over the monster spill in the Gulf of Mexico, the end of the story has not been reached. The spill has

---

1. Now also available online at http://nigeriang.com/money/oil-politics-the-coming-belt-of-fire/3326/. (accessed 15 June 2016)

revealed difficulties in oil field practices even where sophisticated technologies are involved.

The environmental health concerns of the industry have also been brought to question. Do the companies in the sector conduct genuine environment impact studies/analyses for their projects? Do they have adequate oil spill response plans and mechanisms? To what degree were the health and safety of the workers considered in the ill-fated Deepwater Horizon rig in the Gulf of Mexico?

In terms of transparency, we see that the spill volume kept increasing over time, as the BP was forced to be more realistic with the figures. It is a shameful display of corporate duplicity and unwillingness to be open. Check this trend.[2] April 25: 1,000 barrels; April 28: 5,000 barrels; May 27: 12,000 – 25,000 barrels; Early June: 20,000 – 50,000 barrels per day. Today it is generally agreed that the spill was spewing over 100,000 barrels a day right from day one.

Spineless government officials

In our backyard, ExxonMobil has recorded a string of offshore spills from their Qua Iboe operations since last May without a whimper from government about the plight of the local communities and their destroyed fisheries.

The impacts of the spill in the gulf have made headlines and cleaning efforts are even televised. What no one knows is the extent to which these will affect the food chain and ultimately humans. What is not known also are the cracks that the explosion may have caused on the ocean floor and what the implications maybe if there is a huge release of gases like methane from the earth bowels.

The clean up efforts are sustained, but the burning of crude releases greenhouse gases and the use of a cocktail of chemical dispersants pose untold dangers.

Photos of the impacts on birds and aquatic life melt even the stoniest of hearts. Little wonder government officials have attempted to keep them from public view. What breaks my heart more than those photos from the Gulf of Mexico is the nonchalance of our government officials about destroyed livelihoods and destroyed human lives in the Niger Delta, in the Gulf of Guinea.

A 2006 report[3] by Nigerian scientists and the World Conservation Union concludes that 'an estimated 1.5million tons of oil has spilled in the Niger Delta ecosystem over the past 50 years, representing about 50 times the estimated volume spilled in the Exxon Valdez oil spill.' The Exxon Valdez spill occurred

2. See Time Magazine's 100 Days of the BP Spill: A Timeline. http://content.time.com/time/interactive/0,31813,2006455,00.html (accessed 31 May 2016)

3. See Kadafa, Adati Ayuba (2012), Environmental Impacts of Oil Exploration and Exploitation in the Niger Delta of Nigeria (accessed 31 May 2016); and https://globaljournals.org/GJSFR_Volume12/2-Environmental-Impacts-of-Oil-Exploration.pdf (accessed 7 june 2016)

in 1989. Till date clods of crude oil are still traceable on the shores that were impacted. And that spill was cleaned 21 years ago.

When will there be a real response in Nigeria?

27

Gas flaring, hot air and fertilisers

This was first published in 234NEXT[1]
Last week, Goodluck Jonathan signed what has been described as binding memoranda of understanding (MoUs) with petrochemical companies from Saudi Arabia and India as well as with Chevron, AGIP, and Oando. According to the president, this step signalled the start of a gas revolution in Nigeria.

Coming a week before general elections, we cannot fail to note the political undertones in the timing of the launch. Past governments have made pronouncements on their determination to halt the heinous acts of gas flaring over the past decades. These have amounted to nothing but hot air.

Administrative measures to curb the menace started in 1969. Ten years after the initial moves, the 1979 Gas Reinjection decree set 1984 as the essential date when gas flaring became outlawed in Nigeria. However, the penalty for flouting the law was a slap on the wrist to the oil companies so that they continued flaring, poisoning the environment and maiming the people.

The last set dates for ending gas flaring were given by the late Yar'Adua in December 2008. Towards that deadline, Odein Ajumogobia, at that time the minister of state for petroleum, announced that a new flare out formula was being worked out to end gas flaring without hurting government revenue.

When an earlier target date of December 2007 was getting close, the same minister announced that zero gas flare was a moving target.

The gas revolution announced by Mr. Jonathan is replete with figures on how much money would be spent on the various projects, but as far as news reports go, we have seen very little of the volumes of associated gas currently being flared that the projects would take up.

The drums are very loud that foreign direct investments will bring in US$10 billion and an aggregate investment of US$25 billion over the next three years, with activities in fertiliser production, petrochemicals, and methanol manufacturing.

All these will add up to create about half a million jobs directly and

1. It can also be accessed at http://nnimmo.blogspot.com.ng/2011/04/oil-politics-gas-flaring-hot-air-and.html (accessed 31 May 2016)

indirectly. But statistics can be colourful, especially when they are of the Nigerian variety.

Except for Chevron, which says it would start by delivering 175 million cubic feet of gas a day 'once the pipelines and infrastructure are in place'[2], we don't see concrete gas utilisation figures associated with this revolution.

Undoubtedly, efforts have been made in the past by some oil companies to reduce the amount of gas flared. For example, the Nigerian National Petroleum Corporation (NNPC) and Mobil's East Area Natural Gas-to-Liquid (NGL II) project initiated in 2006 was completed ahead of schedule in 2008 and was designed to utilise 950 million standard cubic feet of gas daily.

Chevron also announced that the West African Gas Pipeline project (WAGP) would significantly dent the amount of gas being flared in the oil fields.

It turned out that this was not the case because, according to some estimates, less than 20 per cent of the gas on this pipeline is associated with crude oil production. The bulk of the gas comes from gas fields, rather than oil fields.

As for the oil company AGIP, their notoriety in the area of gas flaring is marked by their seeking to claim carbon credits for utilising some of the gas they have been flaring at Kwale in the face of the fact that the activity has not ceased to be illegal in Nigeria.

The same can be said of Chevron and their claims of the WAGP as well as of other companies such as Pan Ocean, which is making strides towards obtaining carbon credits through this route dotted with ethical and moral questions.

Nigeria's huge gas reserves, easily accessible in new gas fields, have made the stoppage of gas flaring unattractive to an industry that has admittedly taken the act as a routine matter since the 1950s, despite public outcry. Nigeria is said to have proven gas reserves of about 187 trillion cubic feet.

The 2005 estimates by the World Bank indicated that Nigeria flares about 812 billion cubic feet of gas daily. We can argue all we want whether this figure has increased or reduced with the passage of time.

Oil companies sometimes make curious claims about how much reduction they have achieved in their flaring binge. Some have claimed up to 30 per cent reduction, but the reality on the ground has not backed up such claims.

The gas revolution also has an anchor on the stomach, as marked by the proposed fertiliser plants. Obviously, the existing fertiliser plant in Nigeria has not made a significant dent on supply of the product in the country and this has left the field open for above and below board games.

While launching the gas revolution project, the president declared, 'We can only be successful if our actions impact on the common man in Nigeria. The agricultural revolution arising from the fertilizer and blending plants will

2. Brock, Joe (March 24, 2011),. Nigeria unveils 'gas revolution' weeks before polls,Reuters,http://www.reuters.com/article/ozatp-nigeria-gas-idAFJOE72N0K620110324 (accessed on 31 May 2016)

create affordable food for Nigerians and a lot more for export. The LPG agenda will touch the lives of many households, as cheaper and cleaner LPG displaces kerosene. The disposable income that arises from the savings will result in the purchase of more goods and services, boosting GDP.'[3]

Good lecture, Mr. President. However, when it comes to wholesome food provision for the present and in the future, it has been shown that this will come through farmers who cultivate using agro-ecological methods, and will not be dependent on the use of artificial fertilisers that are climate changers and ultimately harm soils and water bodies.

Let the Gas Revolution roll, but let it begin by the release of the figures of associated gas to be used in the project, as well as the schedule for the environmental and other impact assessments for the project.

And, of course, the question remains, Mr. President: when will gas flares be quenched? Do we take that the revolution will begin to snuff some flares out in three years and continue over indeterminate years into the future?

3. Archibong, Elizabeth. March 25, 2011. Gas project to bring in USb foreign investment. http://www.nairaland.com/630992/gas-project-bring-10b-foreign (accessed 31 May 2016)

28

The bush refineries of the Niger Delta

This piece was first published in March 2011 in 234NEXT[1]
The recent public presentation of the book, 'The Ogbunigwe Fame' by Felix Oragwu[2] brought up memories of the technological innovations that kept the Biafran dream alive from 1967 to 1970.

During the vicious civil war, Biafra was blockaded and starved of access to resources ranging from domestic goods to industrial products. Necessity thrust upon Biafra the need to innovate and to create. It was in this mode that the nascent nation built and ran crude oil refineries and also produced missiles or bombs, then known as 'ogbunigwe' or 'Ojukwu buckets'. These efforts were driven by the inescapable urge for survival.

In the past few years, there has been an emergence of what many term 'bush refineries' in the oil fields of the Niger Delta. These are spots in the swamps and creeks where local people, mostly youth, produce petroleum products using crude oil obtained from either already leaking pipelines or from spots broken into by crude oil thieves.

These refineries pose serious health hazards to their operators as they have no clue about the toxic nature of the products and do not have any sort of protective clothes, boots, or gloves. These young folks bear the extreme heat from the flames of the belching dragons in order to produce litres of semi-refined products that pose additional threats to the end users.

Many deaths related to kerosene explosions have been recorded and these may have resulted from the use of the uncontrolled products from these contraptions. The dire poverty in the oil region is often cited as justification for the existence of these bush refineries.

Regrettably, the response from the government, as well as from the political parties seeking control of the federal government after next month's elections, is nothing beyond the provision of physical infrastructures in the

---

1. It can be accessed at http://nigeriang.com/money/oil-politics-the-bush-refineries-of-the-niger-delta/8332/ (accessed 31 May 2016)

2. Scientific and Technological Innovations in Biafra: The Ogbunigwe' Fame 1967-1970. Fourth Dimension Publishers, Nigeria, 2010

region. While these are essential, the most urgent need of the region, and indeed the entire nation, is the detoxification of our environment.

As we have often argued, the average Nigerian will take care of her basic needs if the physical environment supports her livelihood-generation efforts. This means that the urgent first step is an audit of the environmental situation of the region, as could possibly be exemplified by the current study of Ogoni by the United Nations Environmental Programme (UNEP).

A factor that could be perpetuating the bush refineries is the dislocation of the social infrastructure of the region. This includes the loss of communality, the rise of individualism, and the deep corruption that has been entrenched by key players in the oil industry sector. These systemic ruptures must be structurally addressed.

We cannot ignore the efforts of security agencies in combating the menace of the illegal refineries. But merely combat posturing only gives the trigger-happy security men cover for extortion and further human rights abuses of an already traumatised people.

However, it must be acknowledged that the continued operation of these bush refineries is a disservice to the local people and a huge shame to the government.

'Ghost' bush refineries

Going by figures from the Joint Military Taskforce operating in the Niger Delta, hundreds of these bush refineries have been destroyed. By mid-December 2009, the JTF reported that there were over 1,000 'illegal refineries' in the Niger Delta and that within two months to that time they had destroyed 600 of the refineries in different parts of the region.

Sarkin Bello, the General who commanded the JTF at that time, made an important point that just as other ills had started in one part of the nation and spread to other parts, there was a chance that such refineries may pop up in other areas of the country— especially those through which oil pipelines passed.

Months later, Mr. Bello bemoaned the resurgence of the bush refineries, as was widely reported in the mass media. It was not exactly surprising when a fortnight ago, the JTF announced that they had detected 500 bush refineries in the Mbiama area on the border between Rivers and Bayelsa States.

It was not surprising because the refineries have been operating more or less brazenly, with law enforcement agents sometimes accused of exacting tolls or illegal taxes from the operators. So they probably destroyed 600 in 2009[3] and the ghosts of the levelled plants resurrected soon as the security agents left the scene. These bush refineries are huge tourist attractions for foreign journalists and you do not need a space rocket to gain access to their locations.

We have heard some politicians claim that the bush refineries cannot be

3. Nigeria military destroys 600 illegal oil refineries. http://www.panapress.com/Nigerian-military-destroys-600-illegal-oil-refineries--12-531101-30-lang1-index.html (accessed 31 May 2016)

eliminated because the youth cannot find alternative avenues of employment. Quite specious, that form of reasoning. It is illustrative of the ineptitude of persons in power who ought to provide employment and keep people away from practices that are harmful to them, the environment, and the economy.

There are untold dangers related to operating these bush refineries. The poor youth who work these refineries, covered in crude, standing in the searing heat and continually inhaling toxic elements can hardly be in a position to enjoy the fruits of their labour. These refineries may put some kobo in their pockets, but they are essentially condemned to poor health and truncated lives.

It is a shame that a government that trumpets amnesty for people who took up arms against state structures would not consider extending the same largesse to these poor lads who are killing themselves. Could they not benefit from some technical education and other benefits extended to the militants?

A point that we must underscore is the fact that despite the large number of these bush refineries and the fact that they refine products that are illegally obtained, their operations do not lead to a reduction of the crude oil output of Nigeria. Why is this? It is simple to see.

Large-scale illegal bunkering with international dimensions has gone on unchecked for decades and many top guns obviously benefit from it. The large-scale crude oil theft in Nigeria has gone on alongside the continual meeting of the production quota of the nation.

The bush refiners may have been inspired by the fact that between the oil wells and the export terminals is a bottomless pit in which thievery is highly rewarded. Efforts at halting the petty stealing for bush refining will not be successful if the cancer of mass oil theft by the high and mighty is not tackled.

29

Seekers of selective transparency

This blog piece was first published in April 2011[1]

By April 15, 2011, important financial reforms proposed by the US government will come into force.

Transnational corporations listed on the US Securities and Exchange Commission (SEC) have been scrambling to insert clauses that would allow them continue operations on the well-worn path of double 'transparency' standards. Observers hope there will be no serious changes in the SEC rule as proposed.

The reform requires that mining, oil and gas companies who trade their shares on the American stock exchange issue an annual report detailing the 'type and total amount' of payments they make to foreign governments. This order is seen as necessary if Barack Obama's Executive Order of January 18, 2011 on improving 'Regulation and Regulatory Review' is to be complied with.

In separate letters written to the SEC by the American Petroleum Institute and the mining giant, Rio Tinto, these bodies laid out reasons why there should be clauses that offer escape hatches through which they can choose which laws to obey and which to break – the laws of the United States or the laws of the countries in which they operate.

This may sound rather quirky, even murky, but the fact is that there are countries in Africa where domestic laws classify revenues obtained from the extractive sector as 'state secrets'.

On account of this, Rio Tinto argues,

> We believe that there should be an exemption, if such reporting would violate, or may reasonably be deemed to violate, host country laws...The issuer should not be forced to choose between which law it will violate— the US or the host country laws.[2]

1. http://www.nigeriaa2z.com/2011/04/07/oil-politics-seekers-of-selective-transparency/ (accessed 31 May 2016)

2. Letter from Rio Tinto plc to US Securities and Exchange Commission, http://www.sec.gov/comments/s7042-10/s74210-44.pdf No longer available online

The American Petroleum Institute (API), on the other hand, focuses at length on what they see as the 'potential for competitive harm 'of the reform.

Their argument is that disclosures would enable competitors who are not listed on the SEC to use such disclosed figures to undercut or outmanoeuvre them in bids. They also opine that disclosures may endanger the lives of workers in the sector as 'energy companies have already experienced numerous incidents where facilities have been sabotaged, operations disrupted, or employees endangered by those who oppose the host country government or energy development.'

Interestingly, sector players do not have qualms with the Extractive Industries Transparency Initiative (EITI) that requires the private sector to disclose payments made to governments and for the governments to disclose payments received.

According to Rio Tinto, Because the EITI also encompasses disclosure by governments, of payments they receive from companies, we believe it is more effective than the proposed rules at improving governance and eliminating corruption in both the private and public sectors. Therefore, we urge the commission to follow the EITI principles to the fullest extent possible.'

The API in its reaction recommends, 'that the commission require issuers to report payments based upon amount actually paid by the issuer to the government entity (as opposed to the issuer's net share of the payment), consistent with the EITI practices.'

Furthermore, the API does not 'believe it is necessary for the rules to specifically list other types of fees that would be subject to disclosure. We note that fees related to entry into, or retention of, licenses or concessions can be competitively sensitive information.'

Analysts see the requirement of the reform as not just a threat to the companies in the sector who are used to having smooth rides over corrupt waters in certain countries, it should also worry governments in Africa and elsewhere who are not open to disclosures of payments made in the sector. Fingers have been pointed at many resource-rich countries.

For example, according to the API, their members 'can confirm to the commission that disclosure of revenue payments made to foreign governments or companies owned by foreign governments are prohibited for the following countries: Cameroon, China, Qatar, and Angola.'

In this context, we cannot escape noticing that some countries have sought to cover their paths by taking shelter under the EITI processes so readily applauded by the companies.

However, it is noteworthy that this quest is not necessarily a smooth sail for such countries. We have the example of Equatorial Guinea as well as Sao Tome and Principe who were expelled from the EITI process at a review meeting in March 2010. At that time, countries whose candidacies were questioned but not cancelled included: Nigeria, Sierra Leone, Congo Republic, and Democratic Republic of Congo.

Ethiopia's harsh response to civil society and her refusal to allow international NGOs to work on advocacy in the country foreclosed her attempt to even gain candidacy status in the first place. And although Nigeria has since become EITI compliant, key areas of revenue opacities still remain to be resolved and corruption in the sector still moves at a galloping pace.

The extractive sector companies require levels of goodwill of host governments to operate in their countries, mainly because of the huge environmental and human rights abuses that accompany their actions.

Providing escape routes and bending the rules to suit their practices would entrench double standards and deepen the endangerment of the environment and peoples.

30

Charge them with manslaughter

First published on 14 April 2011 by Pambazuka News[1]
People who have suffered the impact of unjust practices and those who have been victims of abuse from corporate impunity will heave a sigh of relief the day directors of such companies are brought to court from behind their corporate shields. The spins and the twists in legal tangos that play out so impassively will become a thing of the past.

Whereas corporations do not sweat in the dock, their directors, who are human like the rest of us, may. It is also possible that pleas from the dock would be couched in humane terms and that actions and reactions would become more or less equal as they usually are in physical matters.

In sum, people would sense that justice is reachable in many cases of confrontation between them and corporate entities.

These are some of the hopes being raised by the possibility of top guns at BP being charged for manslaughter over the Gulf of Mexico oil spill of April 2010. If this happens, it will send a strong signal to leaders of companies that expose their workers to extreme personal risks.

It will also send signals to companies engaged in reckless activities that severely impact people and degrade their environment. In addition, it will offer a glimpse into what may become the norm if an international environment or climate crimes tribunal is set up for cases of ecocide.

It has been reported that investigators are pawing over documents and emails that may indicate whether Tony Hayward, former BP chief executive, and other top management officers made decisions or played key roles in what led to arguably the most horrendous environmental disaster in US history. That incident killed 11 workers and spewed yet unknown barrels of crude oil into the Gulf of Mexico.

The internal investigation carried out by BP immediately after the disaster showed that their managers misread pressure data and authorised workers of the Deep Water Horizon rig to replace drilling fluid in the well with seawater – one of the moves in cost cutting suspected to have triggered the disaster.

1. Now available at http://allafrica.com/stories/201104150796.html (accessed on 31 May 2016)

BP has admitted to having made some mistakes but sticks to the claim that they were not 'grossly' negligent.

There is something quite gross about that word 'gross' before the word 'negligence'. If it sticks, the possible fines to be slapped on BP may rise from about US$5 billion to US$21 billion. It will also complicate things for BP in their dealings with the partners on the rig, as they seek to share the costs of the clean-up expected to reach about US$42 billion.

The significance of this case would also be found in the fact that the directors of BP would be unable to hide behind the corporate shield, as is often the case with corporate entities who are persons before the law only for as much as capacity to earn income is concerned; and are phantoms when it comes to responsibility for acts of impunity.

Think how instructive it would have been to line up the directors of Chevron for the environmental crimes in the Ecuadorian Amazonia or those of Shell, Exxon, Chevron, AGIP and the rest for their human rights and ecocide in the Niger Delta. If manslaughter charges are pressed against officials of BP, then the days of companies only being fined and the directors avoiding the dock will soon become history.

Obviously, BP and other corporations will not take kindly to this move. Their arsenal is loaded with tools with which to frustrate legal procedures. Some of them have batteries of lawyers with whom they harass hapless victims and keep the wheels of legal suits spinning.

There is no need to wonder how corporations have got away with murder all the time. One fact is that governments have over the years become largely privatised in the sense that they depend on corporations for revenue and for monetised solutions to virtually every problem.

While suing directors of companies may be a daunting prospect, considering their propensity to keep cases dragging endlessly, it is nevertheless a necessary step towards giving companies a truly human face and maybe a human heart.

We cannot avoid reaching the conclusion that companies behave in a heartless manner because they are fashioned to be unaccountable and can carry out inhuman acts without blinking an eyelid.

Are you not struck by the fact that oil company leaders are ordinarily nice and personable persons, but that this genial nature changes once they put on their corporate toga?

PART IV

IMPACTS OF EXTRACTIVE ACTIVITIES

31

Death and the kids of Zamfara

First published in 234NEXT in October 2010[1]
Four months ago, news broke of the deaths of 163 children in Zamfara State, Nigeria. Interestingly the cause of death, attributed to lead poisoning, was not ascertained by Nigerian health officials but by an international humanitarian NGO, Medecins Sans Frontières (Doctors without Borders).

Since that announcement we have received reports of the death toll rising to about 400 kids. This is a tragedy of monumental proportions.

So far the responses of government have been twofold: a quick announcement reiterating the banning of illegal mining, and also that the area was being decontaminated. What has been termed illegal mining is actually a demonstration of a lack of seriousness on the critical issue of resource management as well as environmental management and protection. Mining of any sort is a hazardous activity. This includes legalised oil and gas exploitation that grimly sends many Nigerians to untimely graves through pollutions and through violence. This suggests that the issue is more fundamental that the legality or otherwise of the activities.

We are also concerned about claims relating to the decontamination of the environment of the polluted communities. The sort of reported casual announcements give a sense of false security to the hapless local people and also a false impression suggesting the existence of acceptable government action. With years of unregulated artisanal mining in Zamfara State and other mineral rich areas, there is an urgent need for relevant government agencies to conduct serious environmental investigations with a view to mitigating the impacts. Outlawing artisanal mining without provision of employment to the army of the unemployed will neither stop the activity nor detoxify the environment.

The tragic decimation of the children of Dareta Village in Anka LGA and Yar Garma in Bukkuyum LGA must be treated with the seriousness it deserves and steps taken to halt it. It should also be understood that simply closing down artisanal mines does not mean that the environment is no longer toxic. In fact, the impacts being noticed now could have resulted from historical lead

---

1. http://nigeriang.com/money/oil-politics-death-and-the-kids-of-zamfara/4840/ (accessed 31 May 2016)

poisonings in the area. This also suggests that disaster possibly lurks in those poor and neglected communities.

Some community people do not even believe that the deaths are results of lead poisoning or any other fall-out of mining activities. Muazu Marafa, a community spokesperson at Yar Garma, for instance, told environmental monitors in June that they do not believe that lead used in the mining process was responsible for deaths in the community because they had been using it for over many decades. In a nation where post mortems are rare and where people are content to say that their relatives died after a brief illness, we see that much work needs to be done to realign attitudes to the realities of available modern knowledge.

Where are the regulatory agencies?

Besides struggling with the National Agency for Food and Drug Administration and Control (NAFDAC) over who has oversight over what territories, it is essential for the Standard Organisation to take a serious look at an existing threat to public health from further lead poisoning in Nigeria. For one, many countries have phased out leaded petrol and in Nigeria the toxic product is the norm. This means that apart from the visible smoke bellowing from the ancient automobiles on our streets, people are inhaling invisible toxins from even the clean exhaust pipes.

Another sore area that needs the focus of the SON is the unacceptably high level of lead in the paints manufactured, sold and used in Nigeria. A recent study by some non-governmental organisations revealed that Nigerian paints contain levels of lead several times above acceptable limits set by the World Health Organisation and that they rank among the highest levels of lead in paints in the world. The paints tested in the exercise include samples from the biggest multinational paint manufactures in Nigeria. What this means is that the threat of lead poisoning is everywhere in Nigeria, on the streets, in our schools, homes, hospitals, everywhere. We have heard of the death of over 400 children in Zamfara State. It is known that lead can absorbed by ingestion, inhalation, and via the skin. Its impacts range from minor irritations and fatigue to others such as gastrointestinal disturbances, neuromuscular dysfunction, personality changes, cerebral oedema, renal failure, and gout.

How many more kids are on the throes of death? How many more are still being poisoned even today? How about the adults who are more resistant to the poison and so remain alive but have their mental capacities severely compromised? Decontamination of the polluted communities requires more than simply closing the mine pits and carting away top soils from obviously impacted areas. There is urgent need for deeper examination of even the soil strata to ascertain the reach of the elements. The fact that water ponds on which the local people depend are also impacted means an urgent need for safe water supply. Shallow wells will simply spread the deaths further. The communities of Zamfara State require proper pipe borne water supply as life saving measures that go beyond political party logos painted on crumbling walls of community huts. Indeed, with the level of pollution and the deaths recorded and still expected, it would not be a radical idea to relocate the communities to safer locations. No effort should be spared in tackling the lead menace to save the lives of the kids of Zamfara State.

32

Resurrection in Chile

This article was written in October 2010 and published in 234NEXT[1].

The live coverage of the rescue of the 33 miners who were entombed in Chile's copper and gold mine for 69 days captured a global audience. It was one of the few moments when good news eclipsed the bad. It was a celebration of human resilience and a picture of the efforts of humanity to search for resources at extreme locations. Spare a moment to ask how many miners would have emerged alive if such an accident had occurred in your country.

In all the celebrations that followed the rescue, few questions were asked about why the mine collapsed in the first place. Was this a rare occurrence here and elsewhere? It is reported that the San Jose mine was so unsafe in 2007 that it had to be closed down for a while. We note that on 30 July, six days before the mine accident, the Chilean labour department had warned again of 'serious safety deficiencies.'[2] Until the 33 miners got sealed up in the mines, the government is not known to have taken any action.

Official data in Chile shows that 373 workers died in mining accidents in the last decade. In 2010 alone, 31 lives were lost.

The mining sector is Chile's main economic powerhouse. The largely privatised mines reap huge profits. However, fatal mining accidents in this country are as high as 39 every year. As the miners emerged from the tomb, the government lapped up the limelight— who wouldn't— and the applause that resounded across the globe. It was also interesting to see President Evo Morales of Bolivia visiting the mine to meet with the lone Bolivian miner who was among the rescued men[3]. This miner had immigrated to Chile for lack of employment in his home country. President Morales offered the man a promise of a job as well as a house. Hopefully, it will not be a job in a Bolivian mine.

With regards to the San Jose mine, in 2007, there was a complaint filed

1. No longer available online. It was written following the rescue of 33 Chilean miners who had been buried for 69 days

2. John Pilger: 14 October 2010. While the media watches the miners, Chile's ghosts are not being rescued. New Statesman.http://www.newstatesman.com/south-america/2010/10/pilger-chile-pinochet-mapuche (accessed 31 May 2016)

3. See Chile rejoices over problem-free rescue of miners. http://www.independent.co.uk/news/world/americas/chile-rejoices-over-problem-free-rescue-of-miners-2105593.html (accessed 31 May 2016)

at the Chilean appeals court and the National Geology and Mining Service by workers of the company together with unions of other companies following deaths in the mines. At that time, the workers demanded the closure of the mine due to poor mine ventilation and lack of proper escape routes. The mine was shut on 22 September 2007 and reopened in 2008, without any changes in the safety provisions.

Stories of industrial accidents emerge regularly around the oil industry. The oil spills of the Niger Delta are daily in occurrence. The massive sludge spill from an aluminium company in Hungary raised huge safety issues about industrial practices, but was almost eclipsed by the reports of the Chilean rescue efforts. As this piece is being written, reports are emerging of a collapsed mining tunnel in Ecuador where four miners are said to be trapped.

As pictures of the families of the Chilean miners camping at the site ran on television screens and websites, viewers could not pick out the fact that some key players were missing. We are talking about figures such as Alejandro Bohn and Marcelo Kemen, the businessmen owners of the San Esteban mines. They left the mine two days after information was obtained that the miners were alive. They did not return there for over two months.

Mining deaths

Thousands of deaths are recorded annually in mining accidents around the world. Recorded figures run as high as 12,000 deaths of workers in the sector every year. In China alone, 2,631 miners died in 2009, while 200 perished in Sierra Leone. In the USA, 26 fatal accidents were recorded in 2007, and 23 in 2008.

Recent deaths from mine accidents in South Africa are 309 in 1999 while 220 died in 2007. In 2008, the deaths added up to 171, while 165 died in 2009. In the first half of this year, 67 deaths were recorded.[4] A rockfall accident in the Marikana mine killed 6 mine workers.

It is shocking that only 24 countries have ratified the Safety and Health in Mines Convention of the International Labour Organisation (ILO) signed in 1995. Chile has not ratified this instrument.

Some analysts have argued that there is already no need for certain minerals to be mined anymore, as enough of the substance has already been brought out of the mines; an example is gold.

As for crude oil, there is an urgent need for the world to move away from fossil fuels and embrace renewable energy sources. The direct and indirect deaths resulting from mining and utilisation of these products should urge us to pause and think.

The resurrection of the Chilean miners, and their return from the bowels of the earth may receive our applause, but we cannot continue to push our luck

4. See Death in the name of profit: South Africa's mine safety scourge. http://m.polity.org.za/article/death-in-the-name-of-profit-south-africas-mine-safety-scourge-2010-08-16 (accessed 31 May 2016)

with unsafe mines, reckless pursuit of capital, and cheap dispensation of human lives.

33

Caught in the Amazon

This article celebrated the conviction of Chevron by an Ecuadorian court in February 2011 over their extreme oil pollution in that country[1]

The people of the Amazonian region of Ecuador could not have wished for a better Valentine Day's gift as a court there slammed a US$8 billion fine on Chevron for heavily polluting the area through serial oil spills.

Texaco spewed the spills into the fragile ecosystem between 1964 and 1990. You may be wondering why Chevron should be caught in the mess left behind by Texaco. Chevron bought Texaco in 2001, although if you should trace their pedigree, you would find that they were already close relatives.

The case against Chevron was first fought in a New York court, but they succeeded in getting the court to agree that the legitimate place to try the case was Ecuador. And so to Ecuador the case went. Perhaps, the oil mogul banked on wearing the plaintiffs out or it may have considered that they could run over the Ecuadorian legal system and come out unscathed, with clean oily hands.

Well, the plaintiffs were tenacious and the lawyers pressed on with the case. In a statement issued soon after the judgement was delivered, Chevron declared that, 'The Ecuadorian court's judgment is illegitimate and unenforceable,' and that it was 'the product of fraud and is contrary to the legitimate scientific evidence.'[2] And, of course, Chevron plans to appeal.

Interestingly, corporations such as Chevron are always keen to avoid liability even when caught in the act with a smoking gun still in their hands. The oil spills they left behind in Ecuador are as evident today as they were decades ago.

Indeed, anyone who makes a pollution tour of the Sucumbios, the region where these environmental disasters are easily visible, will not have to search before seeing the pools of crude Chevron left behind. There are cases of polluted streams, forests, and farmlands. It is like another Niger Delta across the ocean.

1. http://nigeriang.com/money/oil-politics-caught-in-the-amazon/7780/ (accessed 31 May 2016)
2. See Chevron's press statement, February 14, 2011, on the judgement titled Illegitimate Judgment Against Chevron in Ecuador Lawsuit. https://www.chevron.com/Stories/Illegitimate-Judgment-Against-Chevron-in-Ecuador-Lawsuitaccessed on 31 May 2016

The people of the region suffer the impacts of the pollution on their health through diverse cancers, blood disorders, and other such diseases. The corporation also left behind pipelines that often run above the ground and at places people have to stoop beneath them to get into their homes.

One of the more atrocious acts of corporate 'responsibility' was the alleged practice of using toxic drilling muds to make building blocks for schools in the area. Reports also abound of toxic wastes from oil activities being spread on community roads as a social service. Sad thing is that after the operations were taken over by the national oil company, Petro Ecuador, there does not appear to be significant respect for the environment or the people in the region.

Meanwhile, Chevron is kicking and screaming against the judgment. It argues that they have earlier rulings by US and international courts that they can depend on to make the enforcement of Monday's ruling impossible. Their statement is laced with open threats: 'Chevron does not believe that today's judgment is enforceable in any court that observes the rule of law... Chevron intends to see that the perpetrators of this fraud are held accountable for their misconduct.'

The company has loads of money and plenty of time. The poor indigenous people and the campesinos in the Amazon forest have neither of those. To the poor people, it is a fight for survival. For the oil mogul, it is a struggle to avoid responsibility.

When they bought Texaco, it should have been obvious that they purchased both the assets and the liabilities. And when a case is about environmental pollution, pray how do you hide the evidence of several barrels of crude oil in the open environment? It would take particular credulous judges to avoid the physical evidence that cry for justice even before the peoples speak.

The Ecuadoran Amazon communities initially filed a suit in New York City against Chevron in 1993 for polluting their water and soil and sought a settlement in the sum of 27 billion US dollars. After years of struggle and prevarications, the gavel has come down on the table, and Chevron is screaming blue murder.

The company was even reported to have filed complaints against the plaintiffs' lawyers in the USA before the judgment was delivered. Why are these powerful transnational corporations unable to accept guilt and show some respect for local peoples who suffer the impacts of their massive footprints?

The Ecuadorian situation should be a lesson to those who are plucking up Shell's oil fields in the Niger Delta. Someone will pay, one day, somehow.

34

The cemetery of mangroves

Published 4 July 2012[1]
Two visits outside the heart of Rio de Janeiro, Brazil, marked the high-points of my visit to that city for the infamous Rio+20 summit. The first was on 14 June 2012 with colleagues from the Oilwatch International network and that visit took us to Caxias. This is a community that has had to bear fifty years of toxic assault by petrochemical installations including the Refineria Duque de Caxias (REDUC). This refinery is in the heart of petrochemical factories that dot the Caxias landscape and is the fourth largest supplier of refined petroleum products to the country. Potable water is a problem in this municipality and some folks reportedly rely on untreated water from the refinery. The locals see the petrochemicals, including a proposed new refinery set to become the largest in Latin America, as developments that excluded the participation of the citizens.They bemoan a dearth of health facilities even as they bear the assault of multiple pollutions from the petrochemicals complex.

Men and women of the sea

The second visit was on 17 June as part of the Rio+Toxic tour to Magé. It doubled as a solidarity visit to the struggling community people in the Guanabara Bay area. During the visit we meet with members of Homens e Mulheres do Mar Association (AHOMAR) – Association of Men and Women of the Sea in the Guanabara Bay. That name did not include women initially, but after years of gender struggles the role of the women had to be duly recognised and acknowledged in the name.

This last visit commenced from a point between the head offices of Petrobras, the Brazilian national oil company, and the offices of the Brazilian National Development Bank known to be a major financier of toxic projects in the country. The bank has a budget larger than that of the World Bank and extends its tentacles all over Latin America and deep into Africa. The bank turned 60 years on 20 June and fittingly holds itself up as the flag bearer for green capitalism.

1. After Rio+20, Brazil's Cemetery of Mangroves. http://www.theafricareport.com/rio-de-janeiro.html)accessed 31 May 2016)

Life turned unpredictable for the fisherfolk in the Guanabara Bay since Petrobras constructed its pipelines through the Bay. When an oil spill occurred in 2,000 it increased the challenges faced by the fisherfolk. The footprint of that oil spill is still visible in the Ipiringa area and the destroyed mangrove is yet to recover. Indeed, the locals call the area the 'cemetery of mangroves'.

As much as Petrobras has tried to restore the mangrove, the best result is seen only in photos where mangroves planted in pots are photographed before they wilt, according to local sources.

Our team went through various locations in Mage in the company of members of AHOMAR. A rather uncomfortable aspect was that the leader of AHOMAR, Alexandre Anderson de Sousa, had to travel in a police car as it was considered unsafe for him to travel with us in our bus or by any other means. Since 2009 Alexandre and his family have been under 24/7 police protection under the Human Rights Defenders Program of the government. The officers go with him everywhere, everytime.

Perhaps this level of protection is necessary for Alexandre's safety. It could also be a way of ensuring that his activism is curtailed. I found the presence of the cops rather unnerving. But, as Alexandre said, they are living in difficult times and terrain and their struggle is one of survival. Their struggle has been one of ensuring minimal impacts from petroleum installations as well as resisting expansion of the facilities.

Already some communities have been displaced by pipeline construction and their overall fishing grounds has been reduced to about 12 per cent of the area over the past few years. According to the fisherfolk, about 9,000 families are involved in the struggle.

According to research done by the department of Geography of the University of Rio de Janeiro, since the oil spill occurred the fishing stock has depleted by 80-90 per cent of what was the case in the 1990s. Twelve years after the incident, the stock is yet to return to normal contrary to assurances they had received from Petrobras. They regret that the best fishing grounds are no longer accessible to them but are taken up by oil installations, pipelines and related mega-projects.

In addition, commercial fishing companies use big vessels that destabilise the smaller boats used by the locals. In addition, they complain that they get shot at with automatic weapons at times by private security outfit. The objective of the harassment is to stop them from fishing, according to the locals.

'When Petrobras is accused you can be sure there would be no investigations,' one of the local leaders told us. 'We are being squeezed out of business because we cannot go to the deep seas in our small boats.'

Death and dignity

The bay has literally become a platform for Petrobras and sections are fenced off and cannot be accessed by locals. One leader told us: 'We are resisting because we have no options. We might live or die. Our death may not result from gun shots, but because our livelihoods have been destroyed.' He added, 'We are not seeking to be rich, we just want to live our lives in dignity.'

The reality of the precarious situation of the AHOMAR activists was underscored by the murder of two of their leaders a few days after our visit. They are indeed denied dignity in life and in death. The shocking news reached the world that:

> Almir Nogueira de Amorim and João Luiz Telles Penetra, artisanal fishermen and members of Homens e Mulheres do Mar Association (AHOMAR) went missing after going out to fish on Friday, 22 June 2012. Further, reports of the brutal murders inform, 'Almir's body was found on Sunday, June 24th, tied to their boat, submerged close to the São Lourenço beach in Magé, Rio de Janeiro. The body of João Luiz Telles, Pituca, was found on Monday, June 25th, with hands and feet tied in foetal position, close to the São Gonçalo beach.'[2]

Recalling past incidents, reports have it that in 2009, the men and women of AHOMAR occupied the construction sites of land and sub-sea gas pipelines for transport of Liquid Natural Gas (LNG) and Liquefied Petroleum Gas (LPG), built by a consortium between two contractors: GDK and Oceânica, hired by Petrobras. This construction is directly making artisanal fishing impossible in the Mauá-Magé beach, Guanabara Bay, where the AHOMAR headquarters is located.

> They anchored their boats close to the pipelines and stayed there for 38 days. Since then, the fishermen are suffering constant death threats. That same year, in May, Paulo Santos Souza, formerly in charge of the association's accounting, was brutally beaten in front of his family and killed with five shots in the head. In 2010, another AHOMAR founder, Márcio Amaro, was also murdered at his home, in front of his mother and wife. Both crimes have never been cleared up.[3]

Nigerian gas imported and flared

On the way to 'the cemetery of mangroves' we saw gas pipelines that had an interesting story behind them. Around 2002 when Brazil had an energy crisis due to reduction in levels of water in its hydroelectricity dams, the country

---

2. Friends of the Earth International (June 29, 2012) 'Denouncing the brutal murder of two fishermen from Rio de Janeiro'. http://members.foei.org/en/blog/2012/06/29/denouncing-the-brutal-murder-of-two-fishermen-from-rio-de-janeiro (accessed July 14, 2016)

3. Ibid

began to import liquefied natural gas from Nigeria. With an improvement in the energy situation the importation continues and the excess gas is simply flared. It can be said that Nigeria, the second biggest flarer of natural gas after Russia, flares at two ends of the pipe: in the Niger Delta and in Brazil.

Another similarity with the messy oil fields of Nigeria is that most of the spills are first reported by fisherfolk. The Petrobras spill of 2,000 at Ipiringa is said to have occurred by 1 AM and was discovered by fisherfolk six hours later. The massive spill destroyed a huge swath of mangroves and with it took the bottom off the livelihoods of at least 300 families who used to pick crabs, prawns and other seafoods there.

The toxic tour ended with a standing meeting with the environment secretary of the Magé Municipality. Before that meeting we visited Surui community heavily impacted by an oil pipeline that cuts right through it. Stories of buildings cracked by heavy earth moving machinery during the laying of the pipeline as well as displacement of several families are rife here.

The land acquisition process is quite interesting. According to the locals, Petrobras officials would arrive at your door and offer you a certain amount of money for your property. If you refuse, they leave. But when they come a second time they would inform you that the money they offered has been set aside for you in a special account. In other words, you have no option but to accept their offer. When the officials come a third time, their mission is simple: to evict you from your property.

We are all fisherfolk

The deaths of Almir, João Luiz, Paulo and Márcio must be denounced in the strongest terms. We cannot stand apart from this assault simply because it is not occurring in our territories. Our realities are not different whether in the oil fields of Nigeria and Ecuador, the mines of Philippines or the tar sand pits of Alberta Canada. Communities with oil, gas and mineral resources are daily being assaulted. The least we can do to defend our common humanity is to stand in solidarity with challenged peoples all over the world and proclaim that: we are all fisherfolk; we are all AHOMAR activists!

35

The emperor with no clothes

Written in January 2011 after a Dutch Parliamentary hearing o Shell's ecological footprint in the Niger Delta[1]

The Dutch parliament yesterday placed the Royal Dutch Shell before the mirror in a ground-breaking act of scrutiny over the severe environmental and social footprint of the oil giant on the Niger Delta.

Shell may be the only one being grilled but that does not by any means suggest that the likes of Chevron, Exxon, ENI and Total are not mired in the serial abuses in the region. The spotlight at The Hague needs to be replicated in Washington, Rome, Paris, Oslo, and elsewhere.

The Dutch parliament's action is very significant and illustrates how lawmakers should keep their ears open to the cries of the peoples they represent. It should send a signal to their counterparts in Nigeria who prefer to keep a blind eye to the destructive extractive practices going on in the country.

It is widely acknowledged that Shell's operations in Nigeria fall far short of international standards. They do not only spill huge volumes of crude into the marshlands and creeks of the delta, they have also been stoking the air with toxins and greenhouse gases for decades with no sign that this will stop.

It should be noted that the Dutch parliamentarians are not examining Shell's actions based on mere hearsay; some of them had to come to the Niger Delta to see things for themselves. As has been said, the evidence of the eyes speaks far more than what is merely told and heard. It is also significant that these parliamentarians did not merely visit the area but also spent time with the oil giant, hearing their stories and probably having helicopter rides over the incredibly ravaged area.

That some of the parliamentarians came to the Niger Delta must be seen as an indication of their commitment to seek information that should guide their decisions and positions in the face of warnings that the region is a no-go area and should not be visited by foreigners.

1. http://nigeriang.com/money/oil-politics-the-emperor-with-no-clothes/7216/ (accessed June 2016)

Discovery mission

One of such parliamentarians to come on a fact-finding visit is Ms. Sharon Gesthuizen, of the Socialist Party. She is also the spokesperson of the economic committee.

When she visited in December, we went to Oben, Edo State, with her, community people, and Sunny Ofehe of the Hope for Niger Delta Campaign (HNDC). Our mission was to see a typical gas flare. And we did.

The facility was set up by Shell over 30 years ago and has been noisily belching toxic elements into the atmosphere all this time. But officers of the Joint Military Task Force (JTF) would not allow us to leave the location. They kept us there until almost midnight before letting us off.[2]

The worst part of this illegal restriction of Nigerians and a foreign parliamentarian was that the soldiers refused to notify their superior officers of their actions and instead resorted to a series of threats, literally at gunpoint. Regrettable as that incident was, it helped to underscore the insecurity in the region and the serious curtailment of the freedom of movement of the people.

If there is one thing that oil companies hate, it is being placed in a situation where they have to respond to issues relating to their activities in the oil fields they bestride as conquerors. This is understandable seeing that the world is so dependent on crude oil and national energy security has been equated to overall security of nations.

Indeed, the oil companies hold the ace in international politics and have the ears of players in state houses and can even chew those ears if and when they wish. At their behest, wars are fought and at their behest policies are shaped to ensure that their wishes come through.

The embedded nature of the companies in the seats of power provides them the audacity to ride roughshod over environments and local peoples in the most blatant ways imaginable.

While the Dutch parliament is examining the situation, Friends of the Earth International, Milieudefensie (Friends of the Earth Netherlands) and Amnesty International have filed a complaint against the oil company before the Organisation for Economic Co-operation and Development (OECD) over the company's outrageous claims that oil spills in the Niger Delta are almost entirely due to acts of the local communities.

The complaint was filed with the Dutch National Contact Point to the OECD and brings up questions on the non-transparent, inconsistent and misleading figures that Shell has given with regard to the causes of oil leaks in Nigeria. The complaint pushes the position that Shell's claims are unjust and that the figures are random and are not independently verified.

One must say that this is not the first time that the company has been challenged over serious statistics. They were challenged in the past over related

2. This incident is documented in my book, *To Cook a Continent: Destructive Extractions and the Climate Crisis in Africa*. Oxford, Pambazuka Press, 2012

spills percentages used in advertisements in the United Kingdom. They backed down after the challenge and stopped their advertisements that sought to lay the bulk of the blame on third party actions.

United Nations Environment Programme (UNEP) officials, with regard to their research work in Ogoni, picked up current figures cooked by their propagandists. Whereas UNEP thereafter sought to distance itself from the percentages cooked by Shell, the oil company still insists on referring to UNEP as having validated their position about the victims being the guilty ones.

It is hoped that Shell's day in the dock of the Dutch parliament will help the world to see the danger of having corporations continue with impunity on the ground and then use random figures to attempt to hoodwink the world.

As we watch events unfold, the question must be asked: when will our lawmakers wake up to the environmental and human tragedies in our nation?

36

Chasing tar balls in the Gulf of Mexico

First published in 234NEXT on September 2, 2010[1]

When I headed to the Gulf of Mexico, I had a lot of expectations. Above all, the trip to Louisiana, Mississippi, and Alabama was a quest to see the remains of oil spill that held the attention of the world right from when it erupted on April 20.

One thing that stood out is that there has been a strong wedlock between the oil and fisheries industry in the Gulf of Mexico. Apart from the strong Vietnamese community in Louisiana who work almost exclusively in fisheries, others are cyclic in working in both the oil and fisheries sectors. Many fisher folks shift into the oil sector during off seasons when fishing is not much of an option.

It was, therefore, not very strange to find them taking up jobs as clean-up agents for BP after the gusher erased any hopes of fishing in the short term and raised huge doubts as to when they will hurl their nets into the Gulf once more. Stories of health impacts are rife, with reports of respiratory and skin diseases routinely dismissed by doctors as being caused by exposure to heat while engaged in the clean-up exercises.

Groups such as the Gulf Coast Fund are said to have offered the clean-up workers breathing equipment, but BP disallowed their use and threatened to fire anyone who used the protective gears. Why would BP do that? To present a picture that the exercise of cleaning the crude was harmless and thus lessen their liability was the routine response.

This pattern has created in the minds of some of the people a conviction that they are so tied to the oil industry that they cannot live without it. This relationship, described[2] by LaTosha Brown of the Gulf Coast Fund as incestuous, is a big impediment to building a critical mass of citizens for long-term defence of their environment. Ms. Brown read this sort of perception as counting of pennies, rather than considering the value of life.

At Port Sulphur, I joined a community meeting in a local church with

1. Available at http://www.pambazuka.org/governance/oil-politics-chasing-tar-balls-gulf-mexico (accessed June 15, 2016)
2. In a private conversation in New Orleans

visiting local council officials from North Slope, Alaska, who are considering allowing oil extraction in their area. The Alaskans heard tales of how the Gulf spill decimated the livelihoods of the local people and how they could not return to fishing just yet due to the fear that their business may be permanently harmed if they introduce polluted fish and shrimps into the market.

Suspicion remains

There was strong conviction that although BP and the government swear that the coast is all clear of the spills, the chemical pollution of the Gulf will persist. The suspicion exists that BP is merely trying to avoid liability by telling the public that the Gulf is clean without convincing proof. The people believe that a lot of scientists have been bought over and that laboratory results were viewed with suspicion. They cited an example of the announcement that percentages of the crude oil released into the Gulf had been dispersed, evaporated, or eaten up by microbes. They were referring to reports such as the one produced by US federal and independent scientists.

Who pays the piper?

'Whoever owns the laboratory, owns the science,' one local stated. One of the participants at the meeting was Riki Ott, a marine biologist, fisherwoman, and author of 'Not One Drop', who was embedded in the post Exxon Valdez oil spill struggles and who was in the Gulf of Mexico shortly after the disaster. In a recent article she wrote in the Earth Island Journal[3], she posited that BP's clean-up is more like a cover up and holding the company accountable will require digging for the truth.

The Gulf of Mexico is said to have over 3500 abandoned oil wells and about 4,000 oil and gas platforms in the industrial archipelago. All these continue to pose threats.

Most people I spoke with said that leaving the oil in the soil is the ultimate solution to these sorts of incidents. What they could not agree on was what the economic implication would be. They also agreed that the health of the environment directly affects the health of the people.

But I was determined to see some tar balls on the beaches or on the waters somewhere in the Gulf. I was not going to be deterred, even though it was on the eve of the fifth anniversary of Katrina and it was raining heavily. With an equally determined friend, we drove from New Orleans to Gulfport and to Dauphin Island in Alabama. This island had witnessed a lot of cleaning up actions and a local fisherman assured me I would see some tar balls here.

We got there, stepped out in the heavy downpour, and walked along the beach. A couple of folks were out fishing and one offered me a catch he did

3. Riki Ott. Hide and Leak. http://www.earthisland.org/journal/index.php/eij/article/hide_and_leak/ (accessed 7 June 2016)

not quite fancy. A few folks were enjoying a swim in the rain. We walked the beaches and searched the earth piled against private property by BP's bulldozers. My colleague even dug a hole in the sand with driftwood to see if some crude would pop up. I must say that our search did not yield any tar ball. We drove back drenched to our boxers, but assured that we could not have stayed back from our mission for the day.

Yes, we could not spot any tar balls, but Derrick Evans of Gulfport was quick to remind me that when the hurricanes come, many things hidden beneath the surface may show up.

PART V

OIL, DESPOTISM AND PHILANTHROPIC TOKENISM

37

The price of a vote

This article was written in February 2011 and published in 234NEXT[1]
Whether the voters' registration exercise has ended or not is not the issue many Nigerians are talking about these days. The concerns about that exercise are largely about the huge sums spent on its execution compared to the number of voters actually registered.

The electoral commission informed us that about 60 million Nigerians have been registered to vote in the April 2010 elections. That is not too bad considering that they had a target of about 70 million. What may sour the statistics would be if the cases of multiple registrations were identified, weeded out, and the total number is big enough to reduce the overall number of voters substantially.

Some analysts claim that the electoral commission spent N1, 500 per voter if they registered 60 million. If this number gets whittled down, it would mean that the cost of registering one voter might actually be higher than this estimate.

Some preliminary questions that come to mind are with regard to the actual value of a voter's card. Is it worth N1, 500 or more? Can the value be enhanced by certain factors or is it plain crazy trying to price the card at all? If voting is a right, can you price your right?

The second layers of questions are to do with the reasons why some people engaged in multiple registrations with one person getting caught with as many as four cards! One can only imagine how many times they had their finger prints captured and how they must have laughed at the high tech system that was not networked and thus could be fooled at will. The electoral commission says they will weed out multiple registrations when all captured data are downloaded into their central system. We shall see.

What will happen to those who are still in possession of multiple cards and are far from getting caught? When will they know that they have been weeded out? It is possible that some may even get through to the voting period without being caught at any time? If that happens, what will be the value of their stock

---

1. Now available at http://www.oyibosonline.com/oil-politics-the-price-of-a-vote/ (accessed 15 June 2016)

of cards? Will they choose to sell the cards or would they vote for all candidates and so stand a chance of claiming that they voted for whosoever won?

It is not likely that a voter who risked all to obtain multiple cards would want to use them for fun. It is reasonable to assume that the intention is to make merchandise of the cards and sell to the highest bidder, who would probably not pay the owner to carry out the multiple voting but would simply purchase the cards and find some ways of using them in more reliable ways that would eliminate the treachery that could occur in the voting booth away from watchful eyes.

Although vote buying may be entrenched in Nigeria, it is not a peculiarly Nigerian phenomenon or invention. When one looks back into history, there are several cases where vote buying was entrenched and was openly advertised. Such cases can be found in the history of the United States of America and in several other places.

In 1812 Britain, a certain noble man, George Venables-Vernon, left his son-in-law, 'one sum not exceeding £5,000 towards the purchase of a seat in Parliament.'[2] Office purchases and related practices were eventually halted through a 1883 Corrupt and Illegal Practices Prevention Act.

In the nineteen-century USA, the price of votes were often quoted, even in newspapers. One paper, The Elizabethtown Post, reportedly quoted the price of a vote in Ulster County as being US$25.[3]

Whereas vote selling and buying has transformed into other phenomena in the Western world, such as campaign donations and lobbying, it is still possible to see it in many countries in Africa. In fact, in some African countries, where vote buying does not suffice, an incumbent loser can simply refuse to vacate office. After much haggling, they may decide to share offices with presumed winners and carry on as if nothing happened. Or you may end up with two presidents.

Analysts have seen that the price of a vote could vary even within the same country and the office for which the politician is seeking. For example, where the national legislature is more powerful in terms of determining the direction of the state and the office of the president is merely ceremonial, then the vote for a legislator becomes more costly.

The average cost of a vote for those seeking election to the national assembly in Sao Tome and Principe in their 2006 election was said to be about US$7.10, although in the capital this was five times more costly. The price of a vote for the presidency was slightly more than half of that for the national assembly because the president wields power mainly on issues of foreign affairs and defence. With oil revenue's floodgates opening up, those who have more influence over the economy pay more to garner the needed votes to sit over the pie.

2. Enyclopaedia Londinensis, or Universal Dictionary of Arts, Science and Literature. Volume 18, 2012
3. Eduardo Porter. Nov 6, 2010. The Cost of a Vote Goes Up. New York Times.
http://www.nytimes.com/2010/11/07/opinion/07sun3.html?_r=0 (accessed 7 June 2016)

In Nigeria, the votes can easily be arranged in a hierarchy of prices starting from the vote for a local government councillor to that for the president. What may be a bit tricky to rank would be the price differential between the vote for a senator and that for a governor. The confusion comes from the fact that many former governors forget that they had governed whole states and often angle to represent a third of their states as senators.

However, in terms of which office is more lucrative (via corruption), that of the governor takes the cake, no matter how much salaries and perks the senators legislate for themselves. If people got elected to provide selfless service, vote buying, ballot box snatching (a form of wholesale purchase of votes), and electoral violence would not be the norm. What is the price of your vote?

38

Running from the senate

Written in December 2010, this took a jibe at the business of running for office in Nigeria[1]

These are indeed heady days for politicians. Meanwhile these are perplexing days for other citizens. Election fever is in the air and things are moving in rather interesting ways. Some 'Excellencies' thrown up by the world acclaimed defective electoral processes of 2007 have been shooed away and political equations and formulations are emerging in new twists and turns.

The removal of some governors has been neat and received with spontaneous celebrations of victories delayed. Some have been less straightforward. Take a look at the case of Delta State. There is going to be a rerun. Due to other happenings that have eliminated some of those who ran in 2007, such as death, carpet crossing or perhaps withering away of platforms, the ex-governor was about to possibly contest against himself. But then the codes were redefined and he can now look forward to a fight.

The buildup to the 2011 elections makes headlines daily. Imagine how innovative Nigerians can get in their bid to steal an election— by simply stealing data capturing machines. If I had one of those I could possibly generate enough voters' cards to warrant the creation of several wards in my yard. And then I could market the cards or the election. In the past we have witnessed the grabbing of ballot boxes during election days; the employment of folks with stiff thumbs to thumb the cards and thump the election. Elections reduce unemployment! This election is also the first (correct me if I am wrong) to throw up a regionally endorsed candidate of a particular party to confront a candidate that arguably runs on a national platform.

New ambition, new platform

And what do you say when a minister who spent months brandishing a rebranding or rebranded Nigeria suddenly rebranding herself as a candidate

---

1. http://news2.onlinenigeria.com/news/general/64440-oil-politics-running-from-the-senate.html (accessed 7 June 2016)

for the senate, not on her party ticket but on another platform? Could it have anything to do with expiry dates of party formations, going by NAFDAC rules?

Come on, that is nothing new on these shores. If you are keen to seek elective office, you do not have to be encumbered by party loyalty. If party A does not give you the ticket, you can simply do a moonwalk on the carpet and pick up the ticket of a party you previously opposed. Call it a game or a dance. Call it what you like. What is important is that you grab a ticket and place your face on the ballot paper.

Do not be surprised if Atiku Abubakar, or even Goodluck Jonathan end up running on platforms you have never heard of.

Profitable venture

But what does it matter what party platform you contest on? If you are running for a seat in the Senate or in the House of Representatives, you may well be simply embarking on a most lucrative business venture. Seeing that the folks in the National Assembly have the knife and the yam as far as their personal emoluments are concerned, who would begrudge you if your desire is to say good bye to poverty, and that as quickly as possible, before the national economy goes kaput?

Nigerian law makers may well be termed money makers. But don't you know the money trickles down? We get regaled with stories of senators and representatives who utilise their constituency allowances so judiciously that at least some people in their areas can boast of being given motor cycles, sewing machines and bags of salt. Do you imagine that those constituency projects will not move Nigeria to the next level?

I cannot imagine why the uproar that these folks in Abuja do earn 10 times more than their Ghanaian counterparts. Or that David Mark earns almost 10 times more than Barak Obama. Are we in Ghana or the United States of America?

What if these legislators have gulped N385.80 billion as salaries and benefits since their inauguration and are sure to swallow N515.80 billion by the time they end their four years term? Nigerians have to know that Abuja is an expensive city and these folks have to recover their electoral investment— the bags of salt, the payment of thugs that helps to fight unemployment on election days, the bottles of schnapps and beer. Yes?

I laughed out loud when I read an analysis that says that the money grabbed by the lawmakers in Abuja is enough to set up infrastructure to generate 2,572 megawatts of electricity. I think the analyst does not know if the bellies of these lawmakers are wired and connected to the national grid, we won't need any further infrastructure to generate electricity.

It has also been said that the cash they have cornered is enough to build four brand new refineries and refurbish the four existing ones as an icing on the cake. What do you need more refineries for? Do you want to throw the petroleum products speculators into the poverty line? *Haba²?* Have you paused to imagine how much petroleum products can be obtained if we simply give these political fat cats bottles to collect their golden urine and ship to the Nigerian National Petroleum Company?

Okay, so many eyes are set on the hollow, sorry hallowed, chambers of the national assembly come 2011. I wish you luck and indeed I may vote for you irrespective of what platform you stand on since there are no ideological demarcations and no party programmes to prosecute and defend.

With all the goodies available in National Assembly and considering the fact that you can simply get there with two items on your agenda: grab and sleep, you may assume I am running too. You are right. However, I am running from the Senate.

2. Nigerian word for 'that is incredible!' or 'come on!'

39

Serving the nation in hostile times

This article on electoral violence in Nigeria was written in 2011[1]

The most haunting words I have read on Facebook are the words penned by one of the martyred youth corps members, Ukeoma Aikfavour, who had served in the restive north of Nigeria.[2] Thinking that the worst was over and perhaps a few hours to the mindless termination of his life, he had written these words:

> Na wao! This CPC supporters would hv killed me yesterday, no see threat oooo. Even after forcing underaged voters on me they wanted me to give them the remaining ballot paper to thumb print. Thank God for the police and am happy I could stand for God and my nation. To all corps members who stood despite these threats esp. In the north bravo! Nigeria! Our change has come.

The stories that continue to emerge from the post-election violence reveal the depth of depravity of the urchins, the rascals whose passion for human blood was unleashed on innocent youth corps members and on persons of different ethnic extraction and religious persuasion. The nation can, however, take consolation in the heroic stories of Muslims who protected Christians at a great risk to their lives. Such Nigerians should be applauded for showing immense courage in the face of these acts of barbarism.

If the report of the refusal of the authorities of Yusuf Bala Usman College of Legal and General Studies, Daura, Katsina State, to provide a bus to ferry youth corps members to a secured camp is true, that marks another case of insensitivity, an abhorrent behaviour, that could have exposed these young Nigerians to harm had not a Good Samaritan stepped in to pay for commercial buses to convey them to safety. The repeated inability of our security forces as well as emergency agencies to help at this critical moment illustrates a huge capacity deficit.

---

1. http://nigeriang.com/money/oil-politics-serving-the-nation-in-hostile-times/9070/ (accessed 8 June 2016)

2. Se Aikfavour Ukeoma: The One That Never Returned. https://storify.com/ogala/the-last-words-of-aik-ukeoma (accessed 7 June 2016)

The low level of care and security provided for these youth who served as ad hoc personnel of INEC in the elections leaves much to be desired. These youth were posted to northern Nigeria to serve the nation and not to be hacked, brutally murdered and burnt on account of the electoral process. It is nauseating to note that in Gombe State, one of the earlier flash-points, the mayhem began on the basis of the fact that the margin of victory of the CPC over the PDP was not as wide as they had wished. A similar trend equally emerged in Kaduna, shortly afterwards.

Apparently, for these folks, victory or loss can provide convenient cover for destruction. It is time for all parties to realise that in this season of bloodshed, tears and sorrow, there will be a tomorrow. This is basic wisdom if any party hopes to stay as a platform for national unity, drawing support from across the nation.

The duty and onus of responsibility for quelling the raging flames lie heavily on the shoulders of our president. While it may be rather far-fetched to call for a state of emergency to be declared in northern Nigeria, we cannot overlook the fact that the dire situation warrants mobilisation of military forces to check these violent outbursts before the nation is engulfed in another senseless and unwarranted orgy of destruction.

Looking at the electoral framework in our country, we must be thankful that for a candidate to win the presidential poll, the person must secure at least a quarter of the votes cast in two-thirds of the states of the federation and the Federal Capital Territory, apart from securing a majority of the aggregate votes cast. If victory were to be dependent on just the winning of a majority of the votes cast, then it would have been a different ball game.

The 2011 elections may not be perfect, but they have clearly been seen as a huge improvement over past efforts. Only a person to whom wrong is right and right is wrong will refuse to acknowledge this.

My heart aches each time a new story emerges from the sufferings of the victims of the post election violence. When I recall that I have enjoyed living in and making Edo State my base over the past three decades on account of being posted there for National Youth Service, the essential usefulness of the programme as a unifying force is real to me. Today, I begin to have doubts, and do support the need for the evaluation and redesign of the scheme. I am still pondering what my response would be if any of my children is posted to serve in restive northern Nigeria. I know they would have to decide for themselves, but a parent's duty is also to ensure that evil hordes and their sponsors do not kill the dreams of our children who wish to serve their motherland.

Goodluck Jonathan's call for all to step up to the rebuilding of the nation is apt at this time when the smoke of battle can becloud our sense of purpose. We urge the president to go beyond the promise of compensation for the victims of the post election violence and take steps to honour the youth who lost their lives while providing heroic service to the nation in these hostile times.

40

Oil, despotism and philanthropic tokenism

This article was published in Pambazuka News, 14 March 2011[1]
Equatorial Guinea sits in the heart of Africa and is the fourth highest producer of crude oil in sub-Saharan Africa after Nigeria, Angola, and Sudan. It has reaped huge revenues from crude oil sales since 1995 when commercial export began, although discovery of the product was made in the 1960s. It is one country whose political experience makes the years of brute military rule in Nigeria a mere child's play in comparison.

The current maximum ruler of Equatorial Guinea took over power in a bloody military coup in 1979, eleven years after that country's independence from Spain. At that time, Teodoro Obiang Nguema Mbasogo was a Lieutenant Colonel and his uncle, Francisco Macia Nguema, was the president. He is said to have personally supervised the execution of his uncle by firing squad and has reigned supreme over the country of less than a million people since then.

The nation's GDP of about US$37,900 is many times greater than that of Nigeria. The truth, however, is that the high GDP does not translate to a better life for the people. Since the ascendancy of crude oil as a major income earner, other aspects of the economy, especially production of agricultural produce such as cocoa, have suffered neglect. Does that not remind you of Nigeria?

While looking up on President Nguema, one could not avoid visiting the pages of Wikipedia[2] where a part of the entry on this man reveals the following: 'In July 2003, state-operated radio declared Obiang to be a god who is in permanent contact with the Almighty' and 'can decide to kill without anyone calling him to account and without going to hell.' He personally made similar comments in 1993. Despite these comments, he still claims that he is a devout Catholic and was invited to the Vatican by John Paul II and again by Benedict XVI. Macías had also proclaimed himself a god.'

Standing up to the despot

The president, his family, relatives, and friends are said to own most businesses

---

1. http://www.pambazuka.org/governance/oil-despotism-and-philanthropic-tokenism accessed (8 June 2016)

2. https://simple.wikipedia.org/wiki/Teodoro_Obiang_Nguema_Mbasogo (accessed 7 June 2016)

in the country. With the severe curtailment of freedom in the country, it has come as a vent of fresh air when the writer, Juan Tomas Avila Laurel, called for change and embarked on a hunger strike demanding an end to the despotic reign in his country.

In a letter to Jose Bono Martinez, the president of Spanish parliament, dated 11 February 2011, Mr. Laurel states among other things that,

'Since you believe so deeply in the moral solvency of President Obiang, who has been in power since 1979, we fervently request that you exert some influence and take steps towards the formation of a government of transition; one in which those who have held positions in the last 32 years in Equatorial Guinea must not take any part.

'This is not a political demand, as it might seem to you, but a socially and morally driven one. We cannot continue living under a dictatorship that eats away at our very souls.

'Mr. Bono, all we are asking is that you find asylum in a safe country for Obiang, his son Teodorin, first lady Constancia, and his brothers and cousins, the generals and colonels who maintain this unspeakable regime. We believe that one-third of the money that any one of them has deposited in banks abroad would be enough to support themselves for the rest of their days. The remaining sum has to be returned to the country.'

The letter ends with a painful plea for intervention: 'Mr. Bono, it is not fair for me to put my life in your hands. I will not deny, however, that whatever happens to me will depend in great measure on what you do.'

Gaddafi's oily stand and neo-philanthropists

The events in North Africa and in the Middle East clearly highlight the fact that crude oil has been largely responsible for the entrenchment of crude regimes in the region.

This is particularly visible in Libya where the man who has been in power for over four decades clings on, threatens to cleanse the country of protesters house to house and if necessary blow up the oil and gas fields of the country.

This threat has introduced a new dimension to the volatility of crude oil supply and threatens to push prices to record high. Call him what you like, but Mr. Gaddafi and his cohorts have fed from the feeding bottle of crude oil and taking that from them without a period of weaning is bound to result in the slaughter and tantrums that are the hallmarks of the regime in Tripoli.

A quick look back at the third week of February 2011 shows that as we saw a fine slammed on the oil giant, Chevron, for polluting the Amazonian region of Ecuador, we heard of the company's philanthropic move in the Niger Delta.

The gesture is a clear case of philanthropic tokenism. It appears that Chevron sought to draw attention away from the long-awaited verdict from Ecuador by moving across the Atlantic and displaying a suspect front of compassion in the bloodstained and oil soaked creeks of the Niger Delta. The link and the timing are inescapable.

The company announced with much fanfare a splash of US$50 million, ostensibly to ignite economic development and tackle conflict in the region—of which, it must be said, the company admitted to being a contributor in the past.

The money is being funnelled through the company's Niger Delta Partnership Initiative and the United States Agency for International Development (USAID) and will be spent over the next four years. The thrust will obviously be to generate employment since the oil company hires only a tiny fraction of the millions it has impoverished through the destruction of the creeks, swamps, farmlands and forests that they depend on for their livelihoods through oil spills, gas flares, and the dumping of other toxic wastes.

These are interesting days indeed. Without doubt, crude oil business is not only volatile, but explosive. It is the stuff that oils the machinery of despotism and it is the stuff that blinds the world to the blood that flows on the streets as people fight for liberty.

It is also the stuff that bluffs and seeks to blind us from demanding environmental justice but accepting tokens.

PART VI

SLIPPING ON OIL AND GAS LAWS

41

Nigeria's unacceptable biofuels policy

This was first published on November 17, 2010 in 234NEXT[1]

At the time the barrel price of crude oil shot up, the world began to sing the biofuels song. Biofuels were touted as a replacement for fossil fuels and the answer to poverty and even the climate crisis. They were presented as being both renewable and environment friendly.

Moreover, it was said that they would not compete with food crops in terms of land uptake, as some of them would be grown only on degraded and marginal lands. The idea of biofuels giving fossils fuels a good fight was so widespread that the formation of a 'green' OPEC was proposed.[2]

Research has shown that biofuels are just as harmful to the climate as fossil fuels when factors such as loss of soil carbon and deforestation are computed. It has been proven that the energy output is actually same or less than what it took to cultivate, process, and transport the fuels. Thus, biofuels are not so green.

The reality of the push for biofuels is that they quickly metamorphosed into agrofuels— targeting food crops and pumping foods into machines rather than empty stomachs. The food crisis that hit the world when commodity speculations, conversion of grains into fuels, and other factors drove food prices up, forcing the mantra of agrofuels of the energy saviour of the world to be re-examined.

Lester Brown, of the Earth Policy Institute, warned in 2007, for instance, that the 'grain it takes to fill a 25-gallon (95 litres) with ethanol just once, will feed one person for a whole year.'[3] In the same year, the United Nations special rapporteur on the right to food, Jean Ziegler, described agrofuels as a 'crime against humanity', and called on governments to implement a 5-year moratorium on their production.[4]

The Nigerian biofuel policy has been gazetted as *Nigerian Bio-fuel Policy*

---

1. Now available at http://nigeriang.com/money/oil-politics-nigeria's-unacceptable-biofuels-policy/5610/ (accessed 15 June 2016)

2. Adam Ma'anit. A Green OPEC? http://newint.org/blog/editors/2007/01/28/a-green-opec/ (accessed 7 June 2016)

3. Lester R. Brown. Food or Fuel? http://www.earth-policy.org/mobile/books/fpep/fpepch4 (accessed 7 June 2016)

4. http://www.un.org/apps/news/story.asp?NewsID=24434&#.Vv03-seYdFI (accessed 7 June 2016)

and Incentives No. 72 Vol 94 and is dated June 20, 2007. Let us briefly look at what the wholesale adoption of the agrofuels highway means to Nigeria and the world.

The push for agrofuels has meant a massive uptake of lands for the cultivation of oil palms, corn, cassava, sugar cane, and jatropha, among others. It has translated to land grabs in Africa, loss of lands by pastoralists to jatropha in Africa and India, and slave-like engagement of farmers as mere outgrowers in many parts of the tropical world.

The rush for agrofuels has some benefits, but the benefits have been for agribusiness, and the losers are small scale and family farmers and pastoralists.

In Nigeria, this rush saw cassava as the major target, with large swaths of farmlands being set aside for cassava to be converted into ethanol. Jatropha has also been an attraction with one company allegedly promoting its cultivation in Ogoni land for the production of what they cheekily call Ogoni Oil! In many parts of Northern Nigeria, the best-watered lands, often along rivers, have been grabbed for agrofuels cultivation.

In many cases, communities have been cajoled to give up their lands and become farm hands to big business on the promises of regular income and a better life that often is nothing more than a mirage.

Biofuel policy

The Nigerian Bio-fuel Policy was produced, packaged, and delivered by the Nigerian National Petroleum Corporation (NNPC) without any public participation. It follows the signature pattern of oil sector arrangements where everything is skewed in favour of corporate actors while the environment is opened to nothing except exploitation.

The policy allows for massive tax breaks and all manners of waivers—exempting the operators from taxation, withholding tax and capital gains tax. They are also exempted from paying import duties and other related taxes on the importation and exportation of biofuels in and out of Nigeria. Moreover, for the first 10 years, such companies would not have to pay excise duties and would also not be required to pay value-added tax.

For what is known as the seeding stage, Nigeria is expected to engage in large-scale biofuels importation. This appears to follow the path already well oiled by the NNPC, a path where Nigeria exports crude oil and still depends on imports of petrol to meet our domestic needs. Starting off with massive biofuels imports may be a clever way of not kick starting the use of the fuel but of entrenching the dependence on imports.

The biofuels policy also recommends a most liberal loan system for the industry, with the funds coming from an 'Environmental Degradation Tax' that would probably include fines from gas flares. The policy expects to profit from continued massive environmental degradation in the oilfields of the Niger Delta, rather than taxing polluters and utilising the funds to detoxify the degraded Niger Delta environment. The policy aims to benefit from the crude oil and also from the damage inflicted on the land and the people.

Instead of requiring that the biofuels sector strictly obeys the Nigerian EIA Act of 1992, this policy requires the Federal Ministry of Environment to 'prescribe standards' for the conduct of Environmental Impact Assessment of biofuels projects. It appears the plan is to ensure the subversion of subsisting laws and regulations.

The policy says nothing about the social and other impacts assessments that an industry of this sort requires. The idea is to build up sacred cows, as seen in the oil industry with its jaundiced joint venture arrangements that allow fines and charges (including community development project costs) to be computed as production costs and, therefore, never touch the profits of the oil companies. In addition, it sees local farmers as outgrowers, with no sense of ownership or control in the entire scheme.

The present Nigerian biofuels policy must be repealed and public debate opened over what sort of policy is needed for this sector.

42

Slipping on oil and gas laws

This article was first published in 234NEXT on February 15, 2011[1]
Over the last two years the National Assembly made attempts to enact laws that would bring about needed changes in the oil and gas sector and in the overall socio-economic environment. Somehow, both the Senate and the House of Representatives slipped into deep sleep over the salient issues.

The first bill that comes to mind is the highly talked about Petroleum Industries Bill (PIB). Oil and gas companies operating in Nigeria have generally been happy to continue business as usual, riding on the tracks set by various military dictators who held sway over the powers of state in the past. The PIB, with all its imperfections, attempts to bring some level of sanity into the sector and allows for some form of integration as well as enabling the nation to derive more financial and socio-economic benefits from the sector.

Expectedly, the oil companies have fought the bill. They have openly said that they would not accept any law that is not favourable to them and have often twisted statistics to suggest that Nigeria is attempting to drive them into bankruptcy if the bill is passed into law without being watered down.

Similarly, the government seems to be bending back and doing the donkey work to ensure that the oil companies are happy. Having been in bed together for so long, the necessary social distance needed for serious negotiations between the government and the companies is difficult to create and so they continue with their pillow talk away from public view.

While the oil companies kick and scream over who gets to pocket how much money, the issues that really concern the local communities living in the oil fields were largely overlooked by the PIB. For example, there are no concerns about the impacts of the sector's activities on the environment. Neither did the first draft make any allowance for community consultations and participation.

This writer fully appreciates the difficulties that governments have when it comes to communities. I often recall a conversation I had with a mines and energy minister of another country over serious agitations from communities

1. Now available at http://nigeriang.com/money/oil-politics-slipping-on-oil-and-gas-laws/6994/ (accessed 15 June 2016)

who feared that mining activities in their communities would destroy their livelihood. They demanded a consultation with the government and the government would not agree to hold one because, according to the minister, the national constitution did not say anything about popular consultations and as such the government could not say what it meant, how it should be held, and who would pay for it.

Even when the community folks were ready to hold the consultation at no cost to government and insisted that this was a right under the International Labour Organisation's covenant, the government would not budge. The only promise our meeting left with was that the government would not proceed with the mining projects until a suitable agreement was reached with the affected people.

Consequently, violent conflicts deepened in the area and it does seem that this is the sort of dialogue that some governments would prefer to have. Conflicts in Nigeria have similar roots.

The PIB has the possibility of making environmental and community concerns central in the sector. The environment has been trashed for long enough and there is need for a cease-fire now. And if we like, we can extend an amnesty to the oil companies too.

Sleepy chambers

The Gas Flares Prohibition Bill of 2008 is another critical bill that has been sleeping in the chambers of the House of Representatives. The Senate passed the bill and going by it, gas flaring would have been outlawed again by the end of 2010. Gas flaring has been illegal in Nigeria since 1984 and a High Court sitting in Benin City affirmed in November 2005 that the activity is indeed illegal and a flagrant abuse of human rights.

Shell informed the world about the origins of gas flaring in Nigeria in a document on their website. 'When The Shell Development Company of Nigeria Limited (SPDC) built many of its first production facilities in the 1950s, there was little demand or market for gas in many parts of the world, including Nigeria. So, Associated Gas (AG) was usually burned off safely— a process called flaring. This remained accepted industry practice as SPDC established a major oil operation across the Niger Delta.'[2]

As you can see, this dastardly act goes back five decades! Gas flaring may have been a practice accepted by Shell and their co-travellers in the pursuit of ecocide, we can loudly say that the practice was never celebrated by the suffering people of the oil region. Neither will communities elsewhere in Nigeria accept it if oil is found in their territories.

The gas flare prohibition law for the first time proposes sanctions that should deter the companies from engaging in the destructive activity. Apart

2. Shell. Gas Flaring. http://s06.static-shell.com/content/dam/shell-new/local/country/nga/downloads/pdf/2013bnotes/gas-flaring-2012.pdf (accessed 7 June 2016)

from prison terms proposed, offenders would pay fines equivalent to market value of the flared gas. The bill also proposes that no company should be given any lease for oil and gas exploitation without an accepted gas utilisation plan.

Now the slumber of the House of Representatives over this matter has allowed the 2010 deadline proposed by the bill to slip by. Added to dinner party deadlines set and ignored by past governments, this one has been swept under the carpet and no future deadline is even hoisted to keep hope alive.

Gas flaring is an abuse that cannot be tolerated for any reason. We have allowed it for long enough. We do not need new deadlines. And the farce of presenting projects with regard to existing gas flares for carbon credit under the United Nations Framework Convention must be halted.

The slippery terrain of the oil sector has dulled the outgoing NASS into sleep and given room for continued abuse and pillage. If electioneering will allow governance to proceed, it is not too late in the day for the legislators to rouse from slumber and do the right thing.

43

How about the Petroleum Industry Bill?

This was written in March 2011[1]
The National Assembly's continued delay in passing the Petroleum Industry Bill (PIB) into law is still a source of worry for Nigerians. Both industry watchers and non-governmental groups raise concerns that the delay casts dark shadows over our country's readiness to tackle transparency issues in the extractive sector. Speculations abound that both the executive and the legislative arms of government are under pressure by oil industry players to keep the bill under the carpet until after the April elections. And, thereafter, probably kill it.

Worried observers believe that, following the conclusion of public hearings which ended a while ago, the PIB left the public domain and has since been held on the surgical table by the government, while the oil companies control both the surgical knives and the anaesthetic valves.

The most strident arguments over the bill have come from the oil companies, who have for decades enjoyed the unfettered privilege of calling the shots in the sector. They have enjoyed unrestricted extraction and have gone without accountability, aided by an entrenched architecture of impunity.

Even the extractive industries audits of recent years have not factored environmental destruction and social dislocations into their accounting processes. Without bringing these into the equation, it is impossible to see how the decimated livelihoods of the poor communities and peoples in the oil fields can be redressed.

The major concern of the oil companies has been over the margins of profit the new regimes proposed in the bill would allow them. In seeking to address one of their concerns, the companies impressed on government to turn the existing unincorporated Joint Ventures between the international oil companies (Shell, Total, Mobil, AGIP, Chevron, Mobil) and the Nigerian National Petroleum Corporation (NNPC) into Incorporated Joint Ventures (IJVs). While government agreed to this realignment, the companies seethe over its insistence on retaining control of the venture. They believe that the

---

1. Originally entitled 'Wither the Petroleum Industry Bill'. No longer available online

funding situation would not be different from what currently pertains, since they would not be able to raise funds for the ventures from the capital market.

The delay in passing the bill into law until probably after the forthcoming elections is seen as a clear case of playing politics over this urgent matter. Even non-governmental campaigners for the swift passage of the bill demand 'all political parties to mainstream the debate on the PIB into their political campaigns as a demonstration of their commitment to transparency in the oil and gas sector.' This demand can be understood as a tacit acceptance that the PIB will likely only come into force on the nod of whoever wins the elections. If that is so, it would mean that the earliest date the bill could become law would be sometime after May 29, 2011, after the winners of the election are sworn in. It could also be that what the agitators are demanding is that all politicians should seek the immediate passage of the PIB into law. Which is which?

Reports on how and why the PIB will not see the light of day before May 29 make interesting reading. Some speculate that the National Assembly will simply not act on the bill until after the elections. Others say that even if the Assembly passes the bill, some insiders in the corridors of power will work out plausible reasons the president could offer for not signing the bill into law. The speculators are all casting their readings on the basis that the power of industry is setting booby traps everywhere and squeezing the bill under the carpet. We need to know. Government should urgently speak up on this.

We hear the minister of petroleum waxing positive about the PIB and suggesting that its passage is imminent. Recent reports quote her as saying that oil-producing communities would earn about N1.1 billion as yearly dividend payments from oil revenues, as part of incentives in the bill. It will be great to know the mechanisms that would ensure that such sums trickle down to the people.

She is also quoted as saying, 'There have been so many discussions, modifications and debates by stakeholders in order to ensure a viable legal and regulatory framework for the benefit of, not only the Nigerians, but also for local and international investors.' The government, according to her, 'will continue to engage the National Assembly to ensure passage of the bill as soon as possible.'[2] Nigerians need to know what these modifications are.

One reason why the international oil companies, and possibly even the NNPC, are resisting the passage of the PIB into law is probably the implication of what implementing strict metering of petroleum extraction would mean to them. The bill requires that royalties be paid on production and not merely on what is exported. Oil companies argue against this because a fraction of their production gets lost during transmission to the export terminals. They claim that such losses are owing to oil theft and that paying royalty on production

2. Elisha Bala-Gbogbo. February 22, 2011. Nigeria Says Proposed Law Will Boost Crude Oil Reserves. http://www.bloomberg.com/news/articles/2011-02-22/nigeria-says-proposed-oil-industry-law-will-boost-hydrocarbons-reserves (accessed 7 June 2016)

would place the burden of securing the transmission lines on them, whereas it should be the duty of government.

The issue of metering and the insistence on payment of royalties on production volumes are essential to bringing sanity to the sector. The government should not back down on this. Nigerians deserve to have factual data on how much oil is being extracted daily from our nation and not just how much gets to the official end of the pipelines. The fact that oil is lost/stolen during transmission is well known. We deserve to know how much is being lost. This can only happen if production is strictly metered. If production data were available, simple accounting would reveal how much gets missing. The next questions would be concerning the points of leakage and who benefits from such leakages. The refusal of industry players to allow this simple step in transparency shows the rogue nature of the game and underscores why the people and the environment continue to suffer at their hands.

With all the politicking over the PIB and all the surgery going on outside public purview, who knows what creature will eventually emerge from the theatre?

44

The petroleum bill and last minute legislative contortion

This was published in 234NEXT in May 2011 highlighted the poor environmental and community concerns in the Petroleum Industry Bill[1]

Legislative advocacy can be a double-edged sword, if what you fight for is shrouded in secrecy and all you depend on is the initial draft that was in the public domain. One case in point is the much-expected Petroleum Industry Bill (PIB). The PIB has generated so much interest because the oil and gas sector has been left open to manipulation by political and industry players who made massive gains while the nation got short-changed.

Aside the campaign for the passage of the Freedom of Information bill, the clamour for the passage of the PIB has really captured the attention of many. We all remember the recent public demonstrations of extractive sector transparency campaigners in Abuja, demanding that the national assembly passes the PIB into law before their tenure elapses later on this month.

Some observers have been careful to note that passage of the bill, without public inkling as to what the final contents are, could be quite injurious and on that account it is essential that the public be let in on what has been cooked between the legislators, the petroleum ministry (the executive) and the oil companies.

As May 29 draws close and industry watchers expect that the PIB will be passed into law anytime before then, we have sought to have a peek into what the final document may look like. The best we have been able to see is a document that is yet to be cleaned up, but that gives an indication as to what we may expect.

If you have pointed interest in environmental and social elements of our laws, as some of us do, you can expect a PIB that is not as good as the initial draft that was made public and was subject to many comments and inputs.

A cursory look at the items deleted from the original document by the final draughtsmen gives an indication that the pressures for this watered-down

1. http://business-humanrights.org/en/the-petroleum-bill-and-last-minute-legislative-contortion-nigeria (accessed 7 June 2016)

law came heavily from those who care least about the environment and the communities in whose territory the oil fields happen to be.

At the same time one gets the impression that the lawmakers believe that the concerns of the communities can be fully taken care of by allocating some cash to them. This has always been the bait and is not innovative in the least.

The senate committee recommends the deletion of a section that stipulated that oil companies 'be responsible for any environmental damage, pollution or ecological degradation occurring within the licence or lease area as the result of exploration or production activities, in the case of upstream operators and as a result of any licensed activity in the case of downstream activities.'

The reason for the deletion is that another section provides sufficiently for any 'direct' impacts on the environment. Deleting the section is suspicious, just as we note that environmental degradation is not only caused by 'direct' impacts and polluters should not be allowed to carry on with business as usual under this cover.

The 'final' PIB also rejects the proposal to measure production volumes at wellhead rather than at distribution terminals. This will undoubtedly ensure the opacity of the sector and the reckless thievery it engenders. To add to the profit pile of the oil companies' royalty and tax regimes have been manipulated in their favour.

Another section that has significant deletions is found in the provisions for labour rights. The legislators would not allow anything that protects the rights of workers in the sector and the reason given is that other laws already cover such needs. They pointedly deleted the 'right to freedom of association and effective recognition of the right of collective bargaining.' They also chucked out protection against forced labour or use of underaged persons.

In reality, the restriction of collective bargaining rights (including the sustained casualisation of labour) has been a major area of struggle for labour unionists in the sector.

The legislators also think that it is wrong to create space for the engagement of federal, state and local governments and communities in promoting and ensuring 'peace and development of the petroleum producing areas.' The reason given for this is that the provision is a mere policy statement and 'has no legal binding character.' At another level, the final PIB rejects the idea of incorporating the existing joint ventures and thus promotes the retaining of business as usual.

On the trump card that should silence communities, the PIB seeks to create a Host Communities Fund which would require that operators pay a 'nominal ten percent equity participation in upstream petroleum operations in the Fund as beneficial owners to hold in trust.' This section is presented in such a contorted way that anyone can dance any which way.

Of the total sum held, 80 percent will, from time to time, be allocated for development projects within the communities. The provision here is that it will be of benefit to communities wholly or partially within the lease areas of the oil and gas operators. It is very interesting to note that the benefiting communities will have to demonstrate their direct involvement or exposure to petroleum operation within the licensing area.

How would the direct involvement of the communities be determined? Watch this: they have to collate the number of oil wells, flow stations, oil terminals and power generating plants in their territory. They also have to sum up the length of pipelines that cross their area and also the number of gas flares. If gas flares suddenly become an asset, one wonders why communities are not equally required to count the number of oil spills as well as measure the volumes of oil spilled into their lands, swamps and rivers for the same purpose.

Gas flares?

One would have thought that the final drafters of the PIB knew nothing about gas flares because even the little mention of this illegal activity in the initial draft has been yanked off the 'final' copy.

If the copy of the PIB we have seen is an indication of what we are to expect, it is clear that another opportunity to sanitise the sector is being squandered. It will be a sad day indeed if the current legislators foist a rigged PIB on the nation in the throes of their departure.

45

Many blind spots

This was first published in 234NEXT on September 08, 2010[1]
A major problem with the Nigerian oil industry can be traced to its regulatory mechanisms. While we should assume that such mechanisms could actually help secure efficient operations of the sector, they have led the sector into murkier waters. For a number of years, the Nigerian president doubled as the minister of petroleum. Busy on many fronts, a number of issues must have gone without strict oversight. Because the president was also the minister, the office had more powers assigned to it.

Some experts believe that because of this setting, the minister of petroleum was allowed wide scope for discretion and decision-making powers, without commensurate systems of review and accountability. The sector is disparately regulated, mainly from the Ministry of Environment and that of Petroleum. How coherently these two perform and how their powers overlap or synergise are issues for another day. But there are many areas we ought to worry about. One area of concern is that the oil sector has created some of the most critical environmental and health problems for the Niger Delta and the entire nation.

Who is the governmental watchdog for the Nigerian environment? The answer to that question may seem obvious. Do you say it is the ministry of environment? You would be right. But that would be only to a point.

When we had a Federal Environmental Protection Agency (FEPA) as a subset of the Federal Ministry of Environment, the answer would have been right to a larger extent than it is now. After the demise of FEPA, another agency with a suspiciously long name emerged in 2007. We are talking of the National Environmental Standards and Regulations Enforcement Agency (NESREA).

Let me confess that I had to visit their website to be sure I got that name right! The duties of NESREA, as stated in the Act by which it was set up, are lofty and should build confidence in the agency. However, there are two key areas that raise serious concern. And they are related.

First area of concern is the composition of the governing council of the agency. Article 3 (viii) of the NESREA Act of 2007 specifies a membership

1. http://nigeriang.com/opinion/oil-politics-many-blind-spots/4128/ (accessed 15 June 2016)

slot in the council for a representative of the oil exploratory and production companies in Nigeria.

Why, we ask, is this space created for the oil companies to regulate our Nigerian environment? We note that apart from a slot allowed for the Manufacturers Association of Nigeria (MAN), there is a provision for the Minister of Environment to appoint 'three other persons to represent public interest.' There is no clue in the Act as to who these three would be and on what basis the minister would select them. Would there be representatives of fishers, farmers, or pastoralists? Would there be youth whose future we are already squandering?

We have picked on the objectionable inclusion of the oil corporations in the regulation of our environment because these entities, while baking the petrodollar pie, are also guilty of causing severe damage to the environment and to the psyche of our peoples.

The submission of this writer is that the oil companies should be in the dock and not on the bench in hallowed chambers of environmental and sundry justice. What they have done in the oil communities is nothing short of criminal.

The second issue, which, as already mentioned, relates to the first objection above, is the stipulation of Article 7 (d) of the NESREA Act. This section states that the agency shall 'enforce compliance with regulations on the importation, exportation, production, distribution, storage, sale, use, handling and disposal of hazardous chemicals and waste other than in the oil and gas sector.'

It is clear from the above that a factor has been inserted here to confer a certain status on the oil companies that keeps them away from being regulated by an agency that sets environmental standards in Nigeria and which is supposed to enforce regulations in the land.

With the biggest environmental abuser excluded from the purview of NESREA, the agency must be truly and fully handicapped to play the role it ought to play in regulating the environment. Consider what it would mean if the United States FEPA had no say about how oil companies handle and dispose of chemicals and wastes in the oil and gas sector.

This exclusion from regulation of the oil companies is shocking and scandalous. However, what makes it more objectionable is the fact that these companies, which continue to commit heinous environmental and human rights abuses in the oil fields and communities, are also elevated to the seat of judgment over other lesser polluters of the Nigerian environment.

This is a sad commentary on environmental regulation in Nigeria. It is unacceptable and needs urgent re-examination and correction. A very basic tenet of justice holds that an offender cannot be a judge in his own case.

The unholy wedlock between regulatory agencies and the oil and gas companies is ripe for a divorce. Perhaps, you will tell us that there are other agencies that regulate the oil and gas companies. You could list the Directorate of Petroleum Resources as one. That would make a good joke if you were on a comedy train. The DPR that is unable to tell us how much oil is extracted from the wells and keeps a blind eye or raises hands controlled by political levers cannot take the place of a central environmental regulatory agency.

NESREA needs urgent attention to help close the dangerous gaps created by her blind spots.

46

Nigerian draft Petroleum Industry Bill criminalises communities

Published on August 2012[1]
The draft Petroleum Industry Bill (PIB) submitted by the Nigerian Federal Executive Council to the National Assembly (NASS) has been long expected and drafts have been vigorously contested over the past decade. This is not surprising considering the array of vested interests and corrupt practices the petroleum sector in Nigeria has become well known for.

A polluting sector

The PIB ought to be predicated on the premise that the petroleum resources sector is a highly polluting sector. It should also have the clear understanding that the resources are non-renewable and are thus finite. It is not a resource that will be available or useful in perpetuity. They will either be exhausted or may simply fall out of use. This demands utmost care to ensure socially and environmentally acceptable practices. Acts that are socially and environmentally irredeemably contaminating ought to be shut down for the sake of present and future generations, irrespective of how lucrative they may be. Laws on environmental, social and related impact assessments suggest this.

Oil spills and gas flares should be dealt with as environmental security matters and clear powers to regulate and control activities, punish violators and restore the environment should be identified – and such should be vested on the Ministry of Environment and not on the Petroleum Resources Ministry (by any name). The minister in charge of the petroleum sector should not regulate a sector in which they are active polluters.

Gas flaring

The provisions on the contentious issues of gas flaring leave a lot to be desired. Gas flaring is already illegal in Nigeria (since 1984 and confirmed by a High

---

1. http://nnimmo.blogspot.com.ng/2012/08/petroleum-industry-bill-seeks-to_28.html (accessed 8th June 2016)

223

Court decision in 2005) and the PIB should not legalise illegality. It is wasteful treatment of a resource and harms both the people and the local as well as global environment.The PIB pointedly seeks to abrogate the legislation that outlawed gas flaring and replaces it with nebulous provisions.

Section 201(1) provides that the Minister may permit and penalise gas flaring as deemed fit (by the Minister). Section 277(2) states that the fine for gas flaring shall not be less than the commercial value of the gas. The PIB should clearly state this in Section 201(1) to avoid the Minister lowering the fine to below commercial value. Indeed, the punishment for gas flaring should not be limited to fines but should have weightier consequences considering its criminal nature.

While the draft PIB states that gas flaring should end on 31 December 2012 Section 275 states 'Natural gas shall not be flared or vented after a date ('the flare-out date') to be prescribed by the Minister...).' This contradiction should be eliminated. An illegal act is already illegal and does not need a terminal date. That date ceased from the moment the act became illegal.

Seeking and accepting gifts

The draft as submitted does not come through as decisively focused on stemming the tide of sharp practices. It will amount to a lost opportunity if issues of transparency and accountability are not unambiguously outlined. The section (33) prescribing the powers to receive gifts should be eliminated. Seeking gifts smacks of a lack of sense of ownership of the resources in the first instance. It is also an open door for corrupt activities.

Discretionary powers of the President to award petroleum leases should not be condoned by the PIB. Such powers provide avenues for questionable practices that abort efforts at transparency and due process. Accordingly, Section 191 should be expunged outright.

Provisions for independently verifiable metering of extracted oil and gas should be stipulated in the PIB. A situation where the State does not know actual daily volumes of extracted crude oil and gas makes nonsense of any talks of transparency and feeds corrupt practices of players in the sector and their cohorts. This is the bedrock of the oil thefts that has become a national refrain.

Ownership, control and criminalisation of communities

Another major consideration must be that the politics of oil piths poor and technologically less developed nations against the rich, powerful and technologically advanced nations. There are no niceties in the sector. Thus the PIB should be clear on issues of national sovereignty and keep an understanding that as long as the international oil companies reap more (or comparative) profits in this sector and nation than they do or can do in other places they will not stop investing in Nigeria. Their kicking and screaming are all tactics to continue to reap inordinate profits.

Investing in the technological development of the sector is mandatory if true ownership and control of the resources is to be secured. There is no real ownership without operational control. A reading of the history of the often justifiably cited Norwegian model shows a clear understanding and practice of this. They outlawed gas flaring right from the onset, invested in technological and manpower development and equally determined to proceed on a controlled pace.

True ownership must include that of the communities living within the areas impacted by these activities. Community ownership should be positioned in a way that promotes adequate contribution to the national economy/purse as well as securing protection of the environment and investments. Sections 116-118 providing for Petroleum Host Communities Fund scratches the issue but remains tokenistic and requires deepening. For example it is not acceptable that communities should bear the cost of environmental restoration following incidents (including civil unrest!) in the oil field simply because a member of the community contributed to the incident. This sort of punishment criminalises communities and cannot be accepted.

Section 294(4) equally criminalises local and state governments for acts perceived to have been caused by sabotage. With these levels of government not controlling security outfits it is objectionable that they should be punished for security lapses that may result in sabotage.

Moreover, the deductions made before payment into the fund ensures that only tokens get paid as the oil operators are clearly in charge of the determination of their production costs.

Among other things, the draft PIB is needlessly verbose and repetitive, especially with regard to the organs to be established, and is not gender sensitive. The National Assembly must be ready to do the needed surgery on this piece of legislation.

PART VII

RESISTANCE AS ADVOCACY

47

Resistance as advocacy in the oil fields of Nigeria

This article was written in May 2012[1]

Fossil fuels extraction is extremely destructive to the environment and to the people. Whether crude oil, natural gas, coal or bitumen, their extraction means abuse of the people and the environment. Furthermore, their use means attacks on Mother Earth. Thus, the fossil-driven civilisation is a cannibal civilisation that eats up people.

The direct attacks on people and communities incubate resistance that manifest in different ways and continue to build up. Unfortunately, peaceful resistance to destructive extraction continues to be met with repression and criminalisation.

We see from the example of Ken Saro-Wiwa, martyred leader of the Movement for the Survival of Ogoni People (MOSOP), that resistance can be conducted in a variety of ways. Mass movement building was the path chosen by the Ogoni people and this continues to inspire other peoples who have a clear objective situation that they wish to overturn.

For Ken Saro-Wiwa, cultural revival was an essential tool. He saw the basic need to fight for the dignity of the people and respect for their cultural milieu with tools including drama, poetry and fiction.

Cultural tools are indeed ready vehicles for spreading messages and communicating with wide and diverse audiences. The power of music and poetry as well as other art forms to shape public opinion and cultural direction is well known. For a people impacted by an average of one oil spill per day and with toxic wastes dumped into their environment, resistance is an inescapable route to survival.

In the history of repression of oil field communities in Nigeria, the major offence of the people remains their consistent call for dialogue and repair of the harm visited on them. The response to the people's call for dialogue with Shell at Umuechem in 1990 led to the destruction of a large swatch of the community as well as the murder of several community people. In 1998 the call by dialogue by Ilaje youths in Ondo State of Nigeria received no attention from Chevron

---

1. No longer available online

until the youths occupied the Parabe platform in a peaceful direct action. The response was a commando style attack of the armless youths by the military conveyed in Chevron's helicopters. In the attack on 28 May 1998, two youths were shot dead, others were injured and both the living and the dead were carted into custody.

Women of the Niger Delta remain a formidable, selfless part of the resistance to the environmental degradation and livelihoods decimation by the oil companies in Nigeria. Their involvement in the struggle hinges on the historical heroic stance of Nigerian women. It grew in the women's wing of MOSOP and reached new heights among the Ijaw women who occupied Chevron's flow-stations between 2002 and 2003 and who in 2011 occupied bridges at Edagberi/Betterland (in Ahoada West, Rivers State, Nigeria) to block access of Shell to their facilities.

The demands of the women have remained largely the same: respect and dignity for them and their communities, clean water and basic infrastructure, jobs for their husbands and sons. In utter desperation the women have been forced to deploy what has been termed 'the naked option' – stripping in protest, as the ultimate display if disgust at an industry that ignores the people and the environment and focuses on nothing apart from profit and power.

Although much of what the world hears of the resistance in the oil fields of the Niger Delta has to do with the violent militancy of 2005-2009, the truth is that there has been a consistent resistance through mobilisations against gas flaring, for example, has galvanised signatures from around the world to tackle the menace. Currently thousands of citizens from around the world are signing petitions demanding that Shell cleans up the mess they have piled up in the Niger Delta.

Communities are also forming themselves into networks, eliminating inter-community conflicts and monitoring and reporting incidents in their territories as a key means of environmental defence. Litigations have also been used in efforts to make the oil recalcitrant companies and collaborating State agents and agencies to listen to reason. Such cases have been pursued in courts both in Nigeria and in the home countries of the transnational companies.

48

Shell shrugs off Bonga fine

First published on 19 July 2012[1]

Shell's reaction to the fine announced by the National Oil Spill Detection and Response Agency (NOSDRA) over its Bonga oil spill of December 2011 is in line with the oil companies' stance of avoiding responsibility whenever possible.

We recall the case of the Ijaw Aborigines who took their complaints of years of environmental despoliation by Shell to the National Assembly (NASS) in 2000. The complainants demanded a compensation of US$1.5 billion from Shell and this was granted by the NASS. However, Shell rejected the outcome with almost the same reason given in the Bonga case. Shell claimed then that the NASS did not have the powers to penalise it or ask it to compensate the claimants. Their resistance was premised on a claim that the NASS was not a competent body to impose such a fine.

The Ijaw Aborigines went to the courts and obtained a ruling that the US$1.5 billion should be lodged in an account pending appeal. At the appeal, Shell among other pleadings wanted the court to decide 'whether or not the investigative power or any of the powers of National Assembly under the 1999 Constitution... extends to the exercise of judicial powers and award of damages... Whether or no the Political Resolution of the National Assembly or any of its Committee made pursuant to a petition brought before the National Assembly or any of its committee, or at all, has any legal effect and/or legal consequences whatsoever.'

The appeal was recently decided with the justices agreeing with Shell that the National Assembly had conferred on itself judicial powers only the courts had constitutionally. They also saw the NASS as having breached the Doctrine of Separation of Powers. The oil company must be popping champagne on gaining this reprieve. However, the issue of the damaged environment and livelihoods remains unaddressed and the aggrieved people are still stuck in the mire.

Another case that highlights the way corporations frustrate poor

1. http://nnimmo.blogspot.com.ng/2012/07/shell-shrugs-off-bonga-spill-fine.html (accessed 7 June 2016)

communities is that of the heavily polluted Ejama-Ebubu community of Tai Eleme Local Government Area of Rivers State, Nigeria. The community sued Shell over a spill that occurred in 1970/1 and after a long-drawn and tortuous process won the case in July 2010 with an award of N15.4 billion or US$100 million as compensation for the massive pollution of their community. The case was first filed in 2001. This judgement has not been executed because the powerful oil company can always find a tiny technical point on which to delay or avoid acceptance of responsibility for their actions.

Reacting to their fine on the 2011 Bonga oil spill as announced by NOSDRA, a Shell spokesman reportedly claimed 'We do not believe there is any basis in law for such a fine. Neither do we believe that SNEPCO has committed any infraction of Nigerian law to warrant such a fine.'[2]

Ghana is a new player in the oil sector. However, oil spills and oil company impunity started showing up before their first commercial shipment of crude oil. After three oil spill incidents in 2010 that country's minister of environment set up a committee to review the situation to aid government decision on what steps to take. After due reviews the government slammed a US$35 million fine on the oil company, Kosmos.

Kosmos' response was an outright rejection of the fine with the argument that the fine was 'totally unlawful, unconstitutional, ultra vires and without basis.'[3] Kosmos argued that they couldn't find where the Minister derived the power to fine it. They could not find any such authority under the Ghanaian Constitution or any other law of the country to impose a fine on any person on account of an oil spillage incident.

Kosmos's kicking, screaming and bullying eventually earned it a drastically reduced fine. The company eventually paid US$15 million as a fine for the spills as well as for withholding data information on their operations.

The outlaw nature of the corporations is entrenched by weak regulatory frameworks within which they work. Consider the case of the several attempts in Nigeria to stop gas flaring. The numerous deadlines set for snuffing out the noxious flames have never been respected. And the fines paid for routine gas flaring are both minuscule and suspicious. Attempts by the previous NASS to criminalise the act could not be carried through as only the Senate did any significant work on the issue.

Just when we thought the current executive draft of the Petroleum Industry Bill (PIB) was setting a December 2012 deadline for halting the harmful practice we hear from the Presidency that the copies in the public space are 'fake' drafts of the PIBs. The 'fake drafts' also seek to place the burden of spills attributed to sabotage on local and state governments. Several issues arise from the shifting of burdens. First it would empower the oil companies to make

2. Joe Brock and Camillus Eboh. July 17, 2012. Shell faces USbln fine over Nigeria Bonga Spill. http://www.reuters.com/article/shell-nigeria-fine-idUSL6E8IHLKO20120717 (accessed 7 June 2016)

3. Kosmos Bullies Govt Over ¢400bn Fine. http://www.ghanaweb.com/GhanaHomePage/NewsArchive/Kosmos-Bullies-Govt-Over-400bn-Fine-190709)accessed 7 June 2016)

more strident efforts to attribute spills to sabotage when actually they are results of their faulty equipment and negligence. Secondly, local governments and states do not have control over any of the existing security forces and so cannot be expected to police oil installations. Thirdly, it would further criminalise the victims and leave their environments degraded while the companies and their partner federal government dance to the bank without responsibility for their acts. And more.

A copy of the 'authentic PIB' obtained as this piece was being concluded shows that the provisions on gas flaring are hardly different from the status quo. There is no deadline for ending gas flaring and fines can only be as determined by the minister from time to time. The 'fake' PIB had stated that offenders would pay the commercial value of gas flared as a clear deterrent. With regard to the duty to restore environments polluted by oil spills due to sabotage, the 'authentic PIB' places the burden on local and state governments just as the 'fake PIB' provided. A cursory review of the 'authentic PIB' shows that there may be further watering down from the draft that went to the Federal Executive Council (FEC). But we cannot say for sure until we obtain the 'authentic draft' that went to the FEC.

It is clear that weak regulatory environments do not just happen. They are politically engineered to suit certain players.

We recall that WikiLeaks reports revealed a top Shell official, Ann Pickard, boasting that the company had infiltrated vital government ministries in Nigeria and so had privileged information about the internal workings of government. At another occasion she brashly stated that the company would not accept any new petroleum law that does not suit them and the politicians.

At an Oil and Gas conference in Abuja, Nigeria in February 2010, she stated that Nigeria's crude production had been dwindling since 2005 and that the proposed PIB would worsen the situation. She described the draft PIB that was before the last National Assembly as a 'cumbersome document.' Analysts suspect that this posture may have sparked the numerous doctoring that the PIB drafts have seen over the past years.

An analysis of the Oil and Gas Conference noted that Shell's alarmist position was unfounded. The report informs that Pedro Van Meurs, a world-renowned energy consultant, who was at the conference, dismissed Pickard's alarm as a common past time of major oil companies. He saw Shell's opposition 'as the natural track taken by a company which mandate is chiefly the making of more and more profit for its shareholders.' He added that he had 'been advising governments all over the world for over 40 years and I know that this is a battle whereby the oil company will try to get out of the parliament the highest possible share. So they make loud noise so maybe somebody out there might be listening to them.'[4]

4. NNPC. Why Shell is Against PIB. http://www.nnpcgroup.com/PublicRelations/NNPCinthenews/tabid/92/articleType/ArticleView/articleId/70/Why-Shell-is-Against-PIB--Pedro-Van-Meurs-Global-Energy-ConsultantAs-Total-expresses-faith-in-Nigerian-oil-industry.aspx (accessed 7 June 2016)

These instances of oil companies shrugging off penalties go deeper than the surface. Nations that depend on export of primary resources for revenue are essentially rent collectors as they often depend on external agencies or corporations to exploit resources found in their territories. As rent collectors they have limited control over what the actual operators do in the field as the operators actually present themselves (and are seen) as benefactors of the *rentier* states. And the states in turn are ready to pay scant attention to human and environmental rights abuses perpetuated by these operators. Examples abound in the case of Nigeria where human and environmental rights abuses have been documented continuously over the past decades. It is thus no news when these corporations ignore court orders or blatantly challenge government agencies that attempt to enforce any form of redress.

Companies will keep calling the bluff of Nigeria and other countries to which they pose as benefactors while in reality they are rapists. This will only stop with strengthening of citizens driven democracy, legislative activism and systemic change.

49

Decades of destruction: Shell in Nigeria

This article was written in May 2012[1]

Shell is the foremost operator in the oil and gas sector in Nigeria. Indeed, at the time it got a license to explore and exploit petroleum resources in Nigeria in 1937, the entire Nigerian nation constituted its concession. Over the years, the company has built a solid reputation of being foremost not in the span and breadth of its operations but in the abridgement of rights, including environmental pollution bordering on ecocide. Creeks, rivers and streams are constantly polluted by oil spills from aged pipelines and faulty equipment. Routine gas flares, illegal since 1984, pump toxic elements into the atmosphere, choking and poisoning the impoverished local people.

Spills don't hide: The case of Ogoni

The release of the Assessment of the Environment of Ogoniland by the United Nations Environment Programme (UNEP) on 4 August 2011 marked a crucial turning point in the degradation history of the Niger Delta. The report is a scorecard on Shell's activities in Nigeria and reminds the world particularly about the company's ignoble role not just in the decimation of the Ogoni environment but in the massive human rights abuses in the territory that culminated in the execution of Ken Saro-Wiwa and other Ogoni leaders – Sunday Dobee, Nordu Eawo, Daniel Gbooko, Paul Levera, Felix Nuate, Baribor Bera, Barinem Kiobel and John Kpuine on 10 November 1995.

UNEP affirmed that pollution is widespread and not merely occasional in Ogoniland, reporting that all water bodies in Ogoniland are polluted with hydrocarbons. Hydrocarbons reached groundwater at 41 sites and in one place the groundwater that serves local wells was found to have a layer of up to 8cm of oil on it.

The report also revealed that benzene, a known carcinogen, is found in drinking water at a level 900 times above WHO standards. Benzene was also found in some air samples in the area. Generally, hydrocarbons were found at levels 1,000 times above Nigerian drinking water standards. UNEP warned

---

1. No longer available online

that most of the people in Ogoni have been exposed to chronic oil pollution throughout their lives, with soils polluted with hydrocarbons up to a depth of 5 metres in 49 observed places.

The report also confirmed that Shell failed to meet the minimum requirements of the Environmental Guidelines and Standards for the Petroleum Industries in Nigeria (EGASPIN), failed to operate according to international standards and failed even to meet its own minimum operational standards. These all show that the oil mogul thrives on double standards in their operations in Ogoniland and, it bears saying, all their areas of operation in Nigeria. Shell ought to be sanctioned and its licence revoked for flouting the laws of the land.

Transparency claims vs reality

Shell would like us to believe it has now turned over a new leaf. In its Sustainability Report 2011, Shell's chief executive, Peter Voser, makes the following declaration: 'We believe transparency in our operations helps build trust. In Nigeria, for example, the Shell Petroleum Development Company (SPDC) launched a website in 2011 that enables people to track details of oil spills at its facilities, whether from operations or due to sabotage or theft, and how it deals with them.'[2] But these transparency claims require interrogation.

The major spill that Shell reported in 2011 occurred at their offshore Bonga Floating Production, Storage and Offloading (FPSO) platform. The spill occurred on 20 December 2011 and Shell made eight updates,[3] but provided no definitive independent report on the cause of the incident. The report ought to have been issued after a team of stakeholders including Shell, government agencies and community representatives would have made a Joint Inspection Visit (JIV) to the site of the incident. Till date no such report has been seen in public.

The Bonga FPSO is situated about 120 kilometres offshore and floats on one-kilometre-deep ocean water. The deepwater facility is susceptible to high risks, as ocean waves and other events can easily result in catastrophic incidents— comparable to BP's Macondo field platform that exploded in April 2010 in the Gulf of Mexico.

The Bonga spill occurred while a vessel was being loaded with crude oil. As it happened, the operators were busy pumping crude oil into the ocean rather than into the vessel. Shell deployed chemical dispersants in fighting the spill. There has been no word as to what those chemicals were and what impacts they may have on the ocean ecosystem and the food chain.

Shell claims that 40,000 barrels were dumped into the ocean before they stemmed the flow. With a history of underestimates, that figure is not trusted.

2. Shell. Sustainability Report 2011. http://reports.shell.com/sustainability-report/2011/introduction.html (accessed 7 June 2016)

3. Production resumes at Bonga and EA. http://www.shell.com.ng/home/content/nga/aboutshell/media_centre/news_and_media_releases/2012/bonga_01052012.html (accessed 7 June 2016)

A test case of Shell's transparency claim is the spills at Bodo in Ogoni, which occurred in 2008/2009. While Shell says that a mere 1,640 barrels of crude were spilled, Amnesty International puts the figure at between 103,000 and 311,000 barrels. An expert, Prof Richard Steiner, estimates the volumes of crude spilled at between 250,000 and 350,000 barrels [4] in a lawsuit filed by the local community against Shell, the figure put forward is 600,000 barrels.[5]

Another example of Shell's lack of transparency in Nigeria is the question of how much oil is being extracted daily. Audits by the Nigerian Extractive Industries Transparency Initiative (NEITI) reveal that Shell and other oil operators in Nigeria do not provide the Nigerian State with information as to the actual volume of crude oil or gas pumped out of the wells in the oil fields of the Niger Delta. Thus when Nigeria is said to produce between 2.4 million to 2.6 million barrels of crude oil a day[6], that figure represents the volume of crude oil officially accounted for at the distribution points. What happens between the wells and the distribution points is sheer mystery.

Figures put forward for daily crude oil losses in Nigeria range from 130,000 barrels to 300,000 barrels a day.[7] The highest estimate is one that says that as much as is officially accounted for may actually be stolen on a daily basis.[8] With this level of opacity, it is quaint for Shell to claim any level of transparency in the Nigerian oil and gas sector.

Stopping this yawning black hole should be easy through the installation of meters by which independent measurements can be made, but resistance has been reported:

> There are allegations that high-level official corruption, reportedly involving top government officials and some expatriate oil workers that work in concert with their Nigerian counterparts who compromise themselves for financial gratification, are central to the problem.[9]

4. Prof Steiner quoted in Nnimmo Bassey (2012) *To Cook a Continent – Destructive Extraction and the Climate Crisis in Africa*, Oxford, Pambazuka Press,. p 82

5. The Punch (24 April, 2012). *N'Delta oil spill worse than Shell admits – Amnesty International* http://www.nigerianbestforum.com/ index.php?PHPSESSID=f3ee90497eb1b6362255eb723427583d&topic=162539.0;nowap

6. The Vanguard (22 February 2011) Nigeria's crude oil production rises to 2.6 million barrels Daily – FG. http://www.vanguardngr.com/2011/02/nigeria's-crude-oil-production-rises-to-2-6m-barrels-daily-fg/ (accessed 7 June 2016)

7. Editorial (20 July 2011) *Nigeria's Crude Oil Thefts*. Daily Sun http://www.sunnewsonline.com/ webpages/opinion/editorial/2011/july/20/editorial-20-07-2011-001.html (accessed 7 June 2016)

8. Ojo, Eric (12 November 2009) *Bankole laments illegal oil bunkering in Niger Delta – Challenges security agencies on leakages in public funds*, Business Day, Lagos; Mark Tran (29 July 2004) *Shell fined over reserves scandal*. The Guardian, London http://www.guardian.co.uk/business/2004/jul/29/oilandpetrol.news (accessed 8 June 2016)

9. Non-Metering of Oil Wells. 23 September 2011. http://www.thenigerianvoice.com/nvnews/70585/ 1/non-metering-of-oil-wells.html (accessed 7 June 2016)

Gas flares, carbon pollution and political control

According to the World Bank, gas flaring decreased in 2009 in Nigeria from 21.3 billion cubic metres to 15.2 billion cubic metres. Shell however admits that in 2010 their flares went up 33 percent over their 2009 figure.

Shell claims to have reduced its carbon emissions to 6.1 million tonnes of CO_2 equivalent, and that non–routine flaring at upstream facilities accounted for 35 per cent of their gas flaring in 2011 while the remaining 65 per cent was flared due to lack of equipment to capture the gas produced with oil. They added the untenable claim that the 'Nigeria, where the security situation and lack of government funding has previously slowed progress on projects to capture the gas.'[10] Yet Shell has in the past asserted that they flare gas because doing so became standard industry practice from the early years of oil extraction due to lack of domestic demand for the gas. We note that security concerns do not stop Shell from extracting crude oil. It only stops them from stopping routine gas flaring.

Meanwhile, Shell is actively blocking reform in the oil and gas sector. When the Nigerian government broached the idea of a new oil sector bill, Shell's then Vice-President for Sub-Saharan Africa, Ann Pickard, warned that the oil company would not accept any law that is against the interest of the company. WikiLeaks subsequently revealed that Shell had intelligence to share on militant activities as well as on business competition in the Niger Delta. The leaked cables also revealed that Shell knows how leaky the Nigerian government is. Shell's Pickard is quoted as saying to the US ambassador that 'the GON [government of Nigeria] had forgotten that Shell had seconded people to all the relevant ministries and that Shell consequently had access to everything that was being done in those ministries.'

At the time of writing, a former Shell director sits as the Minister of Petroleum in Nigeria. Shell may not need small fries to snoop and scan pages from that Ministry's bulging filing cabinets. They may not have to rely on low-level officials with tape recorders concealed in pens, tie clips, belt buckles, eyeglasses or cufflinks to record meetings and send transcripts to them. Now they may have copies of whatever document they want forwarded directly as a matter of routine.[11]

Spilling and running

Shell's spilling spree has not let up. At the same time the company is engaging in sales of its oil acreages in the Western Niger Delta area. Oil watchers wonder whether the company may be trying to divert attention from the real issue of the consequences of the environmental degradation that they have caused in the area.

---

10. Shell's Sustainability Report 2011. p 29
11. See *So Shell is everywhere,* Chapter 24, page 147 in this book.

The company has opted to move into the deeper offshore where there would be less oversight so that they can pollute without having to contend with watchful local communities. Some analysts have speculated that companies such as Shell are indicating that they may one day quit the Nigerian fields altogether and they do not wish to be saddled with liabilities.

In the meantime, the sales of the fields will not reduce the central role ofShell in Nigeria's oil fields. Shell owns the crude handling facilities, and so would still be some kind of landlord, standing at the crude evacuation gate and reaping the benefits because the crude handling tariff is a crucial part of an operator's economics. The company is simply moving into a new level of exploitation where smaller companies take the flack while they continue to profit.

The regime of pillage and destruction goes on. Nothing has changed, except the language and layout of Shell's websites. A visit to Nigeria's impacted communities reveals that they are now little more than empty shells of their former selves.

50

Between four farmers and Shell

This article was written in February 2013[1]
The Niger Delta of Nigeria is one of the few territories in the world with a huge reserve of crude oil that is both sweet and easy to reach. The crude found here is called *sweet* because it is of the light variety as opposed to heavy type and has low sulphur content. It is easy to reach because a lot of that oil is onshore and also by the fact that Nigeria can easily be accessed by sea.

The sweetness of the crude has over the past five decades brought bitter experiences to the people of the Niger Delta and the totality of their environment. Oil companies operate here with an audacious level of impunity that cannot easily be comprehended by external observers. The question often asked is why the Nigerian government has not stepped up to defend her people and the environment. A part of the answer is that the hands of government have been tied because they are in joint partnership with the polluting oil companies and share in the financial benefits of the on-going rape.

The struggles of the Niger Delta peoples and communities to secure justice have been going on for several years. It has been a long and arduous struggle. We can mention the revolutionary uprising led by Isaac Adaka Boro[2] in the 1960s, the non-violent struggles by the Movement for the Survival of Ogoni People (MOSOP), the Ijaw Youth Congress (IYC), the Ilaje youths and other nationalities of the Niger Delta.

Litigation has been a veritable tool that has brought mixed results. The Ijaw Aborigines sued Shell Petroleum Development Company Limited (SPDC or Shell) to the National Assembly for decades of environmental degradation. When Shell refused to honour the outcome, the case was taken to the courts in Nigeria[3]. The battle is still on. The Ilaje youths sued Chevron in the courts of San Francisco for human rights infringement committed against them on 28 May 1998 when the oil company flew in troops to attack protesting youths

---

1. No longer available online
2. https://en.wikipedia.org/wiki/Isaac_Adaka_Boro (accessed 7 June 2016)
3. Rhuks Temitope. The Judicial Recognition and Enforcement of the Right to Environment: Differing Perspectives from Nigeria and India. http://nujslawreview.org/wp-content/uploads/2015/02/rhuks.pdf (accessed 7 June 2016)

on Chevron's Parabe offshore platform, killing two and injuring others in the process. That case ended after a stretch of ten years with Chevron acquitted of any wrongdoing.[4] The human rights case over the murder of Ogoni leaders brought against Shell in New York produced a more positive result with the oil company accepting guilt.[5] The same has been witnessed in the case against the company for 2008 and 2009 oil spills in Bodo, Ogoni in the courts in the United Kingdom.[6] Furthermore, a decisive judgement against Shell was obtained in November 2005 in the High Courts in Benin City, Nigeria, when the judge ruled in a case brought by Jonah Gbemre of Iwerekhan community in Delta State, that gas flaring is an illegal, unconstitutional activity and should be stopped.[7] The flares still roar as we write this.

The long route to justice

On 30 January 2013 a Dutch court made a ruling in the case of four farmers against Shell in The Netherlands.[8] The significant success in this case includes the fact that Shell was made to stand in the dock in its home country for environmental offences committed in Nigeria where their subsidiary Shell or SPDC operates. Let us look closely at these cases.

Eric Barizaa from Goi community in Gokana Local Government Area of Rivers State represented his deceased father, Chief Barizaa Dooh. The oil spill in Goi for which the case was brought occurred in 2004. Alali Efanga from Oruma community in Ogbia Local Government Area of Bayelsa State was the plaintiff over the oil spill that occurred in 2006. The plaintiff for another oil spill in Oruma is Fidelis Ayoro Oguru. That spill also occurred in 2006. Elder Friday Akpan from Ikot Ada Udo community in Ikot Abasi Local Government Area of Akwa Ibom State sued Shell for a massive spill that occurred there in 2008. The fifth plaintiff was Milieudefensie/Friends of the Earth Netherlands. Environmental Rights Action/Friends of the Earth Nigeria provided further support on the case.

The defendants in all the cases were the Shell Petroleum Development Company of Nigeria Ltd and the parent company, Royal Dutch Shell Plc.

4. Earthrights International. Bowoto v. Chevron Case Overview. https://www.earthrights.org/legal/bowoto-v-chevron-case-overview (accessed 7 June 2016)

5. John Vidal. 10.06.2009. Shell settlement with Ogoni people stops short of full justice. The Guardian (UK). http://www.theguardian.com/environment/cif-green/2009/jun/09/saro-wiwa-shell (accessed 7 June 2016)

6. Elodie Aba. Shell & the Bodo Community – settlement vs. litigation. http://business-humanrights.org/en/shell-the-bodo-community---settlement-vs-litigation (accessed 7 June 2016)

7. Pollution as a Constitutional Violation: Gbemre's 2019 Case. http://www.scribd.com/doc/21340644/Pollution-as-a-Constitutional-Violation-Gbemre-s-Case-by-Labode-Adegoke#scribd (accessed 7 June 2016)

8. Ivana Sekularac and Anthony Deutsch. January 30, 2013. Dutch court says Shell responsible for Nigerian spills. http://www.reuters.com/article/us-shell-nigeria-lawsuit-idUSBRE90S16X20130130 (accessed 7 June 2016)

The plaintiffs went to court seeking that Shell be held liable for the oil spills in the three communities of the plaintiffs and order the company to maintain its pipelines to guarantee that no more oil spills occur in the future; to clean up the oil pollution in their communities and to pay adequate compensation to the farmers for the damages suffered as a result of the spills.

High points

The case was filed in 2008 and all the preliminary processes including the issue of jurisdiction where determined in favour of the plaintiffs in 2009 with the court in the Hague deciding that it had the jurisdiction to hear the case. The issue of whether the case was the same as the one in Nigeria with reference to Elder Friday Akpan of Ikot Ada Udo was determined in favour of the Plaintiffs in 2010. The court held that the case in The Hague was not the same.

One major setback for the plaintiffs was recorded in 2011 when an application calling on Shell to open its books for inspection and copying by the Plaintiffs was decided in favour of the Defendants. This denial of access to the documents that could have shown the clear links of decision-making processes between SPDC and the Royal Dutch Shell made it impossible for the plaintiffs to prove that aspect of the case. By the ruling of the court, although SPDC is a subsidiary wholly owned by the Royal Dutch Shell, the parent company could not be held liable for their actions.

The full hearing on the cases was held over six hours on 11 October 2012. It was at that hearing that the judgment date of 30 January 2013 was fixed.

The judgment delivered on 30 January 2013 was significant mostly in the fact that for the first time Shell was in the dock in The Netherlands for environmental crimes committed outside the country, in this case in Nigeria. Although Shell could celebrate that they were acquitted of polluting Oruma and Goi Communities, the company was found guilty of polluting Ikot Ada Udoh environment. In fact, observers believe that Shell was acquitted on the other cases simply because the court chose to believe a few grainy videos and photographs produced by Shell as evidence of third party interferences in those incidents. The denial of access to essential documents also gave the parent company the right to draw massive profits from their polluting oil fields in Nigeria while washing their hands of the environmental costs heaped on the communities and the nation at large.

Who sabotages whom?

Significantly the court ruled that Shell was guilty in the case of Ikot Ada Udo because they ought to have taken enough care to avoid their installations being 'sabotaged'. The chorus of 'sabotage' of oil installations has been the song of irresponsible oil companies who use that excuse to avoid liability for rusty and ill-maintained facilities that continue to erupt. This claim has been made attractive by Nigerian laws that absolve oil companies of liabilities where incidents are caused by sabotage. Unfortunately, the definition of sabotage appears to have been equated with just any interference by a third party. Because of this even common thefts from pipelines are labelled acts of sabotage whereas acts of sabotage are of more profound nature and are often political in intent. We note that by the court's ruling even with sabotage it must be shown that it could not have been prevented.

The plaintiffs from Goi and Oruma have signalled their readiness to appeal the judgment of the court. As we said on the day of judgment, finding Shell guilty of the spill at Ikot Ada Udo is commendable but we are waiting to see how Shell can celebrate the faulty conclusion reached by the court that they can be exonerated from the ecocide at Goi and Oruma. Goi in Ogoni is a community that is completely sacked by Shell's pollution and all community people from here have been forced to seek refuge elsewhere as environmental refugees.

Shell's disdain for the wellbeing of communities that suffer the impacts of its reckless exploitation of oil in the Niger Delta has been legendary. The spill at Ikot Ada Udo lasted for months and in open farmland and yet Shell had the temerity to fight to avoid culpability. It is just and fair that it be held accountable for this crime. Shell's reign of double standards must not go unchallenged.

51

Walking on caves of fire

This blog was written in May 2013 after a field visit to coal mines polluted communities in the Witbank area of South Africa[1]

Mining always leaves its footprints in both the sands of time and on the lives of the people and their lives. You may think you have seen it all— especially if you have seen or lived in the horrors of oil activities in the Niger Delta. I thought so too, particularly because I have devoted at least two decades of my life in persistent pursuit of polluted lands (at home and abroad) searching for ways to comprehend the great harm generated by extractive activities.

Some of the places that have left deep impressions in my heart are documented in my book *Oilwatching in South America – Or, Guana Guara – Mudfish Out of Water a Pollution Tour Of Venezuela, Curaçao, Peru & Ecuador.* This book is more or less the diary of a pollution tour of these countries carried out in 1997 under the auspices of Oilwatch International. Others can be found in *To Cook a Continent – Destructive Extraction and Climate Impacts on Africa.*

After many years of following the heavy pollution of communities in South West Durban in South Africa, and with kin ears for developments related to proposed fracking in the Karoo, I was still not prepared for the level of impacts from mining in Witbank, Old Coronation mine and other Highveld communities. This filed trip was organised by groundWork (Friends of the Earth South Africa)[2] as a prelude to Oilwatch Africa[3] conference that was held in Midrand mid May 2013. In the group were activists from eleven African countries.

The field trip in Mpumalanga Province where mines literally turned to walking in minefields! No, we did not rush to the mines. Our first port of call was the offices of the South African Green Revolutionary Council (SAGRC) at Witbank. It was early in the morning, but the comrades were already waiting to receive us. Led by Matthews Hlabane, we were quickly given a short introduction to the Witbank.

1. http://nnimmo.blogspot.com.ng/2013/05/walking-on-caves-of-fire.html (accessed 7 June 2016)
2. http://www.groundwork.org.za (accessed 7 June 2016)
3. http://www.oilwatchafrica.org (accessed 7 June 2016)

Mining started here in 1896 and with it began a reign of land grabbing and pollutions. From the 1950s the environmental problems began to intensify and were glaring and undeniable. Acid mine drainage polluted the water and coal dusts took over the air. With these contaminants it was not a surprise that the locals began to suffer from headaches, dizziness, kidney failures and other diseases.

We were informed that there are eight coal-fired plants in Witbank and up to seven hundred (700) mines from where coal and platinum are dug. But that is not all. There is a pile of 5,000 applications for mining permits, with many of them 'linked to the ruling party,' we were told. Overall, there are 6,000 abandoned mines in the country and among these are the abandoned coal mines of the Highveld.

He regretted that there were no direct gains to the community even though so much 'wealth' was being excavated from beneath their feet. The coal extracted is used for electricity generation and for export. The level of contamination here is so high that an estimated 30 billion Rand will be needed for environmental rehabilitation.

Our visit took us to the abandoned Transvaal and Delagoa Bay Mine (TMDB). On arrival we were greeted by a mountain range of waste and polluted water seeping from the tremendous pile. Walking in this field requires extreme caution. We had to go in a single file, trusting that our guide knew what spot to tread and which could be considered as safe ground. We were bemused and some thought it was preposterous for anyone to insist that we couldn't walk where we pleased. Soon enough we all saw why rebellion was not a good option here. There were cracks in the ground best picked out by trained eyes.

We soon knew we were in the devil's territory when we began to smell sulphur. And then we saw heat waves simmering from holes ahead of us. The smell got stronger as we moved nearer. We were walking over caves of fire. A once luscious land was now 880 hectares of hell!

We were told of, and shown sinkholes scattered in the fields. Any place could crack up any time and a yelp may be the only goodbye to be heard before the victims disappear into the netherworld. These mines are located between two Townships and kids and others traverse these burning mines daily either to school or to work. Some kids are said to have fallen into these sinkholes. And someone hazarded that criminals may also have used these burning pits as convenient places to bury their crimes.

Spontaneous fires started in the mines in the 1930s and they were eventually closed in the 1950s. Interesting. It is said that the fires in the mines were burning both the roof supporting pillars and the roofs themselves. We guess that before the mines were closed, perhaps while one portion of the mine was burning, miners were pressed to keep digging in other parts. That can be understood in an apartheid context. But why are the flames not extinguished and the land remediated today?

Our friends told us that because of lack of adequate public response to their complaints about the air quality and other pollutants, they have had to train themselves on how to manage for themselves. In fact, we were told of occasions when officials brought testing equipment and the community folks were the ones who showed the officials how the equipment was operated. Talk of community empowerment! Tests show that some of the water bodies here are either very acidic or highly alkaline.

Leaving the field of horror, we passed by the Vanchem Ltd factory. Our comrades asked us to look up at the sky. Thick smoke bellowed from the stacks. That was not surprising. But they asked us to note that no birds were flying in the area. Well, that was true. 'They simply die if they try,' we were told. Okay. Get me out of here!

We were told that to keep healthy, workers in this factory are compelled to drink milk everyday. I could not laugh. I have personally heard at an environmental health workers workshop of oil company workers (machine operators) in Nigeria who are urged to drink milk as a way of keeping their bodies purified of pollutants. This myth has also been heard of in India. Workers are kept in the dark hope that milk eliminates the impacts of pollution. See Chapter 17 The 'milking' of oil workers in this book for more about this and the cynical actions of corporations.

Our next port of call was the Old Coronation Township sitting on Old Coronation coal mines. The ground here is very unstable. We were taken to a huge pit into which a preschool disappeared after the ground gave way in 2012. Sinkholes started happening here more than five decades ago.

Many residents of this township 'mine' coal in huge waste heaps in the neighbourhood. Stories abound of kids and women who met their death here when the pile of waste collapsed on them as they dug for the carbon needed for cooking and for heating their shacks.

It was one story of woe after another. We saw women and kids digging for occasional lumps of coal. We heard of resource and job opportunities conflicts with migrant workers from the SADC region. We saw extensive acid/water ponds, devoid of life.

'The graves in Highveld are full,' one comrade tells us. 'If you live here and drink the water, there is a 70 per cent chance you will end up with liver problems.' Sadly, kids sometimes swim in the warm ponds and there is a chance that they gulp in the lethal water. There is a high incidence of sinusitis, asthma, tuberculosis and other diseases.

'The doctors work with the mines and the mines work with the government. The people are left to fall through the cracks. The Highveld is a compost,' another comrade insists.

We were thoroughly depressed at this point. Getting to watch a youth drama group perform was hopefully going to be a relief. Soon we were gathered in a community hall built and donated by a mining company!

Speeches and tales of woe from various cities, townships and communities, the Mpumalanga Youth Against Climate Change drama group took centre stage.

The acting was excellent and the storyline and message were clear and direct. Global warming was better termed 'global burning' and humans were shown as anointed to be the most foolish species on earth. The youngsters declared, 'our governments have failed us, but we will not fail ourselves.'

As we left these heavily polluted communities, Comrade Matthew declared that the Witbank is the most polluted city in the world. A Nigerian comrade retorted that the Niger Delta was the most polluted region on earth. An argument ensued but was happily settled that one was a city and the other a region. But best of all, we ought to be arguing about which is the cleanest and safest, not which is most abused by capital. Would either of these places ever return to health?

52

Ogoni and the agony of a delayed clean up

This blog piece was written in February 2013 almost two years after the release of the UNEP report on the assessment of Ogoni environment. It was only then that there were strong signals that the clean up may actually commence[1]

When the UNEP report on the assessment of the Ogoni environment was released in August 2011 the world was astounded at the level of devastation visited on the territory by decades of oil extraction and pollution.

Ogoniland in Nigeria shot into international glare in the early 1990s when the people peacefully demanded an end to reckless despoliation of their land and waters. When the UNEP report was released there was a general sense of relief that at last a definitive study has been carried out in at least a part of the Niger Delta and that remediation steps would be taken to rescue the people from the impacts of the pollution.

Shell Petroleum Development Company (SPDC or Shell), the major polluter in the territory, paid for the study in a rather poetic turn of events, on the polluter-pays basis. If that was not an admission of culpability in the ecocide in Ogoniland, you may have to invent another word for the crime.

The report showed a staggering level of pollution that would require 25-30 years of clean-up activities if there were to be a chance of real remediation. Many people expected the government to declare Ogoniland a disaster zone. The Ogoni people waited to see some clean-up action. The Nigerian people waited to see some clean-up action. The international community waited to see some clean-up action. That the expected action was not forthcoming was a scandal of massive proportions.

Nothing was done until twelve full months rolled by. In other words, since the report was issued till that date, a full year was added to the estimated time needed to restore the Ogoni environment. But what was done after one year?

It took one year after it had been ascertained that there was no safe drinking water in Ogoniland and that the land itself was polluted to depths of up to five metres in places, for any whisper to be heard from the corridors of power.

1. http://nnimmo.blogspot.com.ng/2013/02/ogoni-and-agony-of-delayed-clean-up.html (accessed 7 June 2016)

The UNEP report set out simple emergency actions to be taken to ensure an acceptable clean up of Ogoniland. One of the key recommendations was that government should set up an *Ogoniland Environmental Restoration Authority*. This authority was to have a starting fund of US$1 billion. Rather than set up this body that would set about the restoration of Ogoni land, what government did was to set up what it calls the Hydrocarbon Pollution Restoration Project (HYPREP). This project has succeeded in planting some pollution warning signposts in Ogoniland and billboards on oil thefts in Port Harcourt.

A cursory comparison of the recommended body and the entity that government created shows that something is critically wrong. Why set up a body that would restore rather than clean up pollution? Ogoniland is badly polluted as it is, to set up a body to compound the pollution is alarming, not amusing. Ken Saro-Wiwa, a stickler for correct concepts and sentences, would have written copiously on this twisted contraption if the jackboots had not wickedly truncated his life in 1995.

UNEP officials led by Erik Solheim, former Norwegian Minister of Environment and International Development and UNEP Special Envoy for Disasters and Conflicts, visited Nigeria early February 2013 to meet with government officials and some partners in Abuja and Port Harcourt. The purpose of the visit was to get a sense of what was being done with the UNEP Environmental Assessment of Ogoniland and to know what the next steps would be.

It is not clear what the team came away with, but we at least know that UNEP is committed to seeing the report implemented and Ogoniland cleaned.

In a statement issued by UNEP at the start of the visit, Solheim who led the team said 'With regard to Ogoniland, the UN system is committed to supporting the government throughout the entire process of implementing the recommendations of the report. On behalf of UNEP, I look forward to coordinated and collaborative action with our Nigerian and international partners in addressing pollution in Ogoniland.'[2]

The Ogoni people are one of the most mobilised peoples anywhere in the world. The umbrella Movement for the Survival of Ogoni (MOSOP) enjoys a high level of support across the Ogoni kingdoms, has provided consistent leadership over the years and is well respected by the people. That is, despite some difficult moments, as would be expected of any serious movement.

The degree of cohesion of the Ogoni people provides an excellent template for government to set about the clean up of the territory in a transparent and easy manner. If there are to be difficulties it should be of the technical kind, not the socio-political varieties.

It is not too late for the government to scrap HYPREP and set up the

2. UNEP. 5 February 2013. UNEP team in Nigeria to discuss steps needed to implement Ogoniland Report. http://www.unep.org/Documents.Multilingual/ Default.asp?DocumentID=2704&ArticleID=9386&l=en (accessed 7 June 2016)

recommended *Ogoniland Environmental Restoration Authority*.We will call this the Authority for short.

HYPREP was a hasty creation to tell the world that at least one step had been taken, one year after the release of the UNEP report.

Government should not be shy to do the right thing. Steps taken in the wrong direction may be many, but keeping in that direction may not eventually lead to the right destination. It is equally wasteful to insist on building on a faulty foundation.

Scrap HYPREP, set up the Authority. This Authority would then set about consulting the people, call mass meetings of the Ogoni people, circulate the popular (pidgin English) version of the summary of the UNEP report which can be downloaded from the UNEP website, present the strategy for the clean up to the people and transparently set out the budget outlay for the exercise. The Authority would have the Ogoni people endorse its broad plan and strategies for implementation and monitoring as well. The Authority should be domiciled in either the Ministry of Environment or in the Presidency. It should by no means be located in the Ministry of Petroleum Resources— a key polluter, through the Nigerian National Petroleum Corporation, in Ogoniland.

The extent of pollution and the need to ensure that the clean up is not an occasion for jobbers must be stressed. As UNEP acknowledges, the clean-up required will be complex and there may not be a single method of getting this done. Any delay means further reducing the quality of life and the life expectancy of the people that has already dropped to just over 40 years mainly due to the hydrocarbon pollution. Bloodshed and great sacrifices have been borne by the Ogoni people. The clean up of the territory is not an occasion for gambling.

The selection of consultants, contractors and the handling of the budget require very strict oversight. While we agree that it is possible to have officials in the Authority to handle the procurement and budgetary matters, it is believed that while the in-house crew play roles in those tasks, an agency such as UNEP should play major oversight roles. If this recommendation were accepted UNEP would not handle any of the clean up jobs, but would play a monitoring role.

We are yet to see the Senate and the House of Representatives taking up the clean up of Ogoniland as a critical issue of concern. They need to. It is their duty to ensure that a proper Authority is set up and that there is adequate budgetary outlay for the tasks with both government and Shell putting the money on the table and having an umpire like the UNEP empowered to warehouse the funds.

Getting things on the right track is extremely urgent. As UNEP stated, Continued delay in the implementation of the recommendations will not only undermine the livelihoods of the Ogoni communities, but will also cause the pollution footprint to expand. In the long run, the findings of the study itself will become dated, and therefore further assessments will be needed, causing additional delays.' UNEP hoped to 'convey this sense of urgency to the stakeholders during' the mission.[3] It would be another scandal if this sense of urgency gets ignored.

3. ibid

53

Two years after the UNEP report: Ogoni still groans

This article was published in August 2013[1]

Two whole years after the United Nations Environment Programme (UNEP) issued a damning assessment of the Ogoni environment, the Ogoni people are forced to continue wallowing in the toxic broth that their lands and waters have been made to become. Ogoniland was once a land that supported productive farming, fishing and related activities. That was so up till the moment the oil-rigs began to puncture holes in the land and crude oil began to be spilled on lands, forests and rivers. The air was clean but that changed when gas flares belched like dragons out for the kill. Today, twenty years after Shell got excommunicated from Ogoni, thick hydrocarbon fumes from sundry pollution hang in the air.

From the late 1980s, the Ogoni people raised alarm over the wholesale destruction of their environment. With careful and robustly peaceful organising. With the Ogoni Bill of Rights of 1990[2] they catalogued their demands for environmental, socio-economic and political justice. Although the Bill of Rights was presented to the Nigerian government, until now there has not been a whisper by way of response to, or engagement with, the document.

The Bill of Rights became an organising document for the Ogoni people and also eventually inspired other ethnic nationalities in the Niger Delta to produce similar charters as a peaceful way of prodding the government into dialogue and action. The Bill noted that although crude oil had been extracted from Ogoniland from 1958 they had received NOTHING in return. We reproduce articles 15-18 of the Bill to illustrate some of the complaints of the people:

1. *That the search for oil has caused severe land and food shortages in*

1. http://saharareporters.com/2013/08/01/two-years-after-unep-report-ogoni-groans-nnimmo-bassey (accessed 7 June 2016)

2. MOSOP-Ogoni Bill of Rights. https://www.mosop.org.ng/publications/35-documents/496-the-ogoni-bill-of-rights (accessed 7 June 2016)

Ogoni— one of the most densely populated areas of Africa (average: 1,500
per square mile; national average: 300 per square mile.)
2. *That neglectful environmental pollution laws and sub-standard*
inspection techniques of the Federal authorities have led to the complete
degradation of the Ogoni environment, turning our homeland into an
ecological disaster.
3. *That the Ogoni people lack education, health and other social facilities.*
4. *That it is intolerable that one of the richest areas of Nigeria should*
wallow in abject poverty and destitution.

This Bill of Rights was the precursor to the Kaiama Declaration of the Ijaws, Ogoni Bill of Rights, lkwerre Rescue Charter, Aklaka Declaration for the Egi, the Urhobo Economic Summit Resolution, Oron Bill of Rights and other demands of peoples' organisations in the Niger Delta.

The UNEP report presented to the President of the Federal Republic of Nigeria on 4 August 2011 completely confirmed the claims of the Ogoni people 'That neglectful environmental pollution laws and sub-standard inspection techniques of the Federal authorities have led to the complete degradation of the Ogoni environment, turning our homeland into an ecological disaster.'[3]

The report found that, without exception, all the water bodies in Ogoni were polluted by the activities of oil companies— Shell Petroleum Development Company (Shell) and the Nigerian National Petroleum Corporation (NNPC). Indeed the report stated that some of what the people took as potable water had carcinogens, such as benzene, up to 900 times above World Health Organisation standards. The report also revealed that at some places in Ogoniland, the soil is polluted with hydrocarbons to a depth of five metres.

The UNEP report revealed that the Ogoni homeland had indeed been turned into an 'ecological disaster,' as the Bill of Rights asserted. We remind ourselves that the UNEP report made recommendations that most of us saw as low hanging fruits that government could easily have responded to assuage the pains of the people and commence a process of restoring the territory to an acceptable state. The apparent inaction is nothing but a squandering of opportunities to rescue a people.

A total clean up of Ogoni land will take a life time or about thirty years at least. That is the length of time UNEP estimates it would require to clean up the water bodies in the territory. And it would require an additional five (5) years to clean up the land. How is that a lifetime? Well, life expectancy in the Niger Delta stands at approximately forty-one years.

At the eve of the first anniversary of the presentation of the UNEP report, the Federal Government hurriedly cobbled together an outfit incongruously named *Hydrocarbons Pollution Restoration Project* (HYPREP). The project was set up basically to hoodwink the Ogoni people into thinking that action was

3. See full report here: http://www.unep.org/disastersandconflicts/CountryOperations/Nigeria/ EnvironmentalAssessmentofOgonilandreport/tabid/54419/Default.aspx (accessed 7 June 2016)

being taken to implement the UNEP report. A year after the setting up of HYPREP under the Ministry of Petroleum Resources— a major polluter of Ogoni land— the only visible acts of implementation of the UNEP report has been the planting of sign posts at some places informing the people that their environment is contaminated and that they should keep off. You could almost laugh, but this is sad and serious. Keep off your environment! No options given. The people still drink the polluted waters and farm the polluted lands. Seafood is still being scrounged from the polluted waters and community people still process their foods in the crude-coated creeks.

Two years after the UNEP report, we believe that it is not too late for the government to act. President Jonathan can:

- Declare Ogoni land an ecological disaster zone and invest resources to tackle the deep environmental disaster here.
- Urgently provide potable drinking water across Ogoni land
- Commission an assessment of the entire Niger Delta environment. An assessment or audit of the environment of the entire nation should equally be on the cards urgently.
- Those found guilty of crimes against the people and the environment should be brought to justice and made to pay for their misdeeds. Blame for oil thefts must go beyond the diversionary focus on the minuscule volumes taken up by bush refiners. The major crude oil stealing mafias must be uncovered. Crude oil and gas volumes must also be metered as demanded by groups such as the Environmental Rights Action (ERA).
- Engage in dialogue with the Ogoni people as to the time-scale and scope of actions to be taken to restore the environment. Issues raised in the Ogoni Bills of Rights and the UNEP report provide good bases for dialogue. Extend this all over the Niger Delta.
- Ensure that the actions to tackle the ecological disaster that the Niger Delta has become are not seen as opportunities for patronage or jobs for the boys. UNEP should play a key oversight role, to ensure quality and to build confidence in the process.
- The body to tackle the problem should be domiciled in the Ministry of Environment and should not by any means be under the polluting Petroleum Resources Ministry.
- Shell should be ordered to urgently dismantle whatever remains of its facilities in Ogoni land along with toxic wastes they dumped in the territory.
- Shell should also be required to replace the Trans Niger Delta pipeline that carries crude oil from other parts of the region across Ogoni territory.
- Clean up the polluted lands and waters.

These are just some of the steps that must be taken urgently. The UNEP report gives a good list of several things that need to be done. The time has come to halt the ostrich posture and to face the national environmental challenges squarely. Two years is long enough. Our peoples have patiently lined up to fall into early graves. Twenty-three years ago several Ogoni people were sacrificed because they dared to speak up concerning the state of their homeland.

A stanza of the Nigerian National Anthem urges, 'The Labours of our heroes past shall never be in vain.' We cannot continue to sing those lines mindlessly while the ecological disaster persists and our heroes groan in their graves.

PART VIII

AFTERWORD

54

Afterword

What has changed since these essays were published? Quite a bit. There have been more oil discoveries in Africa and the scope for environmental pollution is increasing apace. At the same time, the scope for action by citizens to defend their environments is also expanding. New networks of resistance to the destructive impacts of extractives are being born or are being consolidated in Nigeria, Ghana, Uganda, South Africa, Togo and globally. Communities are sharing knowledge through these networks and training on ways of uncovering the sugar-coated lies of the polluters.

Of importance is that there are real indications that a clean-up of Ogoniland will finally start, as called for by the UNEP report on the assessment of that environment. While Ogoni will be the starting point, the clean up will extend to other parts of the heavily degraded Niger Delta and,hopefully, to other parts of Nigeria. We hope that such clean-up actions that contribute to reviving local ecosystems and economies, might spread to every nook and cranny of Africa. It is high time that we stop arguing over which community should claim the title of being the most polluted place in Africa. Instead we should be looking at which is the most pristine and life supporting.

The popular resistance we are seeing includes an expanding corps of community ecological defenders, bold litigations in the face of great odds and increasing legislative actions. The task now is to connect the energies erupting from the disparate arena of struggles in order to construct an unstoppable movement towards to emancipation of our peoples. The environment connects all humans and all living things on the planet. Actions for justice carried out together cannot be stopped.

The chapters on climate change negotiations included the essay on *Ambition, Selfishness and Climate Action* referring to COP18 that was held in Doha in 2012. We quoted UNEP report that said 'the aggregate voluntary emissions reductions by rich, industrialised and polluting nations would not ensure the level of reduction needed to avoid catastrophic global warming.' It showed that a gap existed between the 'level of ambition that is needed and what is expected as a result of the pledges.' At the time of publication of this book we have had three more COPs. The COP in Lima locked in the era of voluntary emissions reductions, while COP21 in Paris in 2015 finally endorsed the voluntary regime together with the submission of Intended Nationally

Determined Contributions (INDCs) to emissions-reduction by nation states. The fears expressed by UNEP years earlier have been confirmed by projections based on the intended emissions reductions proposed by nations.

A pre-COP21 synthesis report prepared by the Secretariat of the UNFCCC on the aggregate effect of the INDCs, which was communicated by parties to the Convention on 1 October 2015, stated:

> If Parties were to not enhance mitigation action until 2030 beyond the action envisaged in the INDCs, the possibility of keeping the temperature increase below 2 deg C still remains. However, the scenarios in the IPCC AR5 indicate that this could be achieved only at substantially higher annual emission reduction rates and cost compared with the least- cost scenarios that start today or in 2020. Therefore, much greater emission reductions effort than those associated with the INDCs will be required in the period after 2025 and 2030 to hold the temperature rise below 2 deg C above pre-industrial levels.[1]

The parties to the UNFCCC know that the INDCs will not solve the climate crisis. A deadly disease requires radical, not convenient, action. Climate negotiations continue to ignore the call for a transition from dependence on fossil fuels. This is a clear indication of their concern for the lives of future generations. It underscores the vice-grip of the fossil-fuel industry.

There is a need to transform our approach to *system change not climate change*. System Change is more than mere sloganeering.The environmental horrors that the planet and its life forms increasingly faces have systemic roots. Radical changes to the system, not polite negotiations, are urgently needed. In other words, system change cannot be negotiated. It will be enforced by people taking directaction. If people fail, Mother Earth will not. She will go on without us.

July 2016

---

1. UNFCCC. 30 October 2015. Synthesis report on the aggregate effect of the intended nationally determined contributions. http://unfccc.int/resource/docs/2015/cop21/eng/07.pdf (accessed 7 June 2016)

About the author

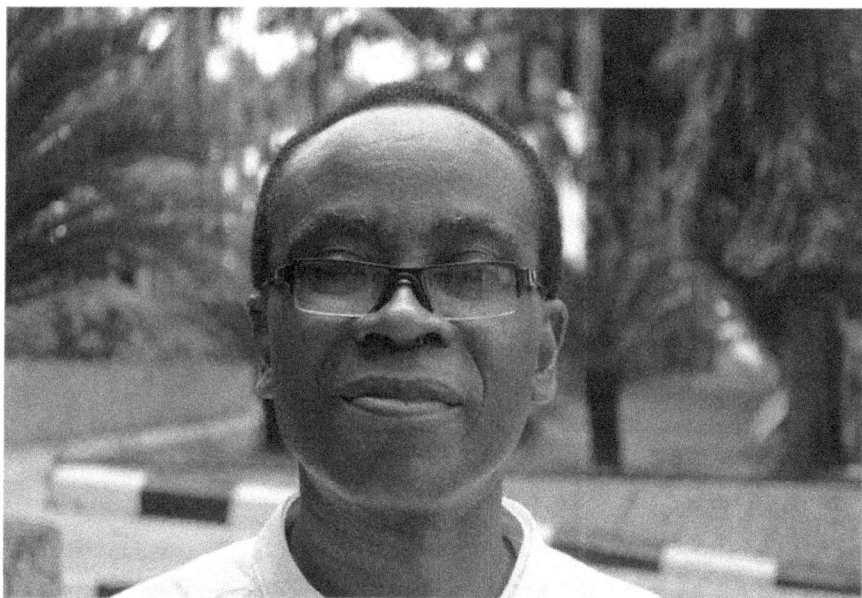

Nnimmo Bassey is a Nigerian environmental justice activist, architect, essayist and poet. He is the director of the ecological think-tank, *Health of Mother Earth Foundation* (HOMEF) and coordinator of *Oilwatch International*. He was the chair of *Friends of the Earth International* (the largest grassroots environmental organisation in the world) from 2008-2012 as well as the co-founder and executive director of *Environmental Rights Action* (1993-2013) which is based in Nigeria (in Benin city, Lagos, Abuja, Port Harcourt and Yenagoa).

He was a co-recipient of the 2010 *Right Livelihood Award* also known as the 'Alternative Nobel Prize.' In 2012 he received the *Rafto Human Rights Award*. In 2014 he was awarded Nigeria's national honour as a *Member of the Federal Republic* (MFR) in recognition of his environmental activism.

Nnimmo Bassey is the author of the highly acclaimed book, *To Cook a Continent: Destructive Extraction and Climate Crisis in Africa* (Pambazuka Press) and, in Portuguese, *Cozinhar Um Continente: A Extração Destrutiva e a Crise Climática na África* (Daraja Press) which detail the destructive impacts of the extractive industries and the climate crises in Africa. He is also co-author, with the No REDD in Africa Network, of *Stop the Continent Grab and the REDD-ification of Africa* (Daraja Press). He has also authored books on architecture.

His poetry focuses on environmental justice. *We thought it was oil but it was blood* and *I will not dance to your beat* are two of his most widely known books of poems.